The Myth of
Scientific Public Policy

SOCIAL
PHILOSOPHY
& POLICY CENTER

The Myth of
Scientific Public Policy

Robert Formaini

transaction

Transaction Publishers
New Brunswick (USA) and London (UK)

Published by the Social Philosophy and Policy Center and by Transaction Publishers 1990.
Second Printing, 1990.

Library of Congress Cataloging-in-Publication Data.

Formaini, Robert, 1945-
    The myth of scientific public policy/Robert Formaini.
        p.   cm. --(Studies in social philosophy & policy; no. 13)
    Includes bibliographical references.
    ISBN 0-88738-352-1. -- ISBN 0-88738-852-3 (pbk.)
    1. Policy sciences.   2. Science and state--United States.
I. Title.   II. Series.
H97.F68   1990
320'.6--dc20   90-30315
                                                                          CIP

Cover Design: Kent Lytle

For my parents, who taught me right from wrong—and that there were no shortcuts for understanding either.

Series Editor: Ellen Frankel Paul
Series Managing Editor: Dan Greenberg

The Social Philosophy and Policy Center, founded in 1981, is an interdisciplinary research institution whose principal mission is the examination of public policy issues from a philosophical perspective. In pursuit of this objective, the Center supports the work of scholars in the fields of political science, philosophy, law, and economics. In addition to the book series, the Center hosts scholarly conferences and edits an interdisciplinary professional journal, *Social Philosophy & Policy*. For further information on the Center, write to the: Social Philosophy and Policy Center, Bowling Green State University, Bowling Green, OH 43403.

Table of Contents

Acknowledgments

Any project which takes six years to complete involves many individuals. I cannot cite all those who have helped me complete this book, but some measure of gratitude needs to be expressed to many who played large roles in seeing that this book was finished.

I would like to thank Dr. John Sommer of the National Science Foundation (Policy Research and Analysis) for his help and suggestions over many years, including his advice that I investigate these issues. Though I have cursed him for convincing me of the wisdom of this course of action, I have profited greatly from the experience.

Dr. Larry Redlinger, Dean of General Studies at the University of Texas at Dallas, played the role of sympathetic critic and dissertation chairman. His suggestions on the reworking of many parts of this book, in both method and structure, all made it better than it otherwise would have been.

Dr. William Seeger of Amber University provided much-needed encouragement, criticism, suggestions, and intellectual stimulation.

The National Science Foundation, through Sigma Xi (the national honor society of scientists), provided two opportunities for the author both to poll and then to personally interview many leading scientists on issues relevant to this book.

The Institute for Humane Studies (George Mason University) provided a Claude Lambe Fellowship and the Ludwig von Mises Institute (Auburn University) provided scholarship support for dissertation-related expenses.

The Social Philosophy and Policy Center's (Bowling Green University) Dr. Ellen Paul (Deputy Director) and Dan Greenberg (Managing Editor) are responsible for the publication of this manuscript and its being much more coherent than it might otherwise have been.

The usual caveat about the responsibility for errors remaining within this work applies.

Introduction

Public policy in the United States is debated, analyzed, and implemented within a framework characterized by the acceptance, explicitly or implicitly, of certain assumptions. One of the main assumptions is the objective nature of the reality which surrounds us, along with a subsidiary assumption concerning the ability of our techniques accurately to explore and to control that reality. It is on this foundation that we confidently craft our public policies. Yet it may be that our confidence is not a function of the objective reality we seek to model, but is rather the result of our having accepted the pronouncements of philosophers, scientists, consultants, policy analysts, and others who have succeeded in convincing most people of the efficacy of their methods of analysis.[1] It may surprise the reader to discover that this intellectual foundation is not everywhere as strong as interested parties claim.

The overall conclusion reached in this volume can be stated quite simply: scientifically-based (i.e., *justified*) public policy, a dream that has grown ever larger since the Enlightenment and that, perhaps, has reached its apogee toward the close of our own century, is a myth, a theoretical illusion. It exists in our minds, our analyses, and our methods only because we seek to find it and, typically, we tend to find that which we seek.[2]

Americans, therefore, demand and are supplied with the "answers" to a variety of complex questions that have emerged from our particular way of life. The muse that we have come to rely upon—to an extent heretofore unprecedented in history—is Science. Occasionally, the muse admits (albeit usually not directly to the public) that it does not have the answers that are being requested.[3] This does not deter us, however, from demanding even more—and more complex—answers from the same source. Concomitantly with our greatest technological triumphs occur our deepest, most profound fears of their by-products. The entire process can be seen, as though played out in farce, through study of the struggle over the Shoreham nuclear power plant. In a time of increasing concerns about air pollution and the Greenhouse Effect, a new 850-megawatt nuclear plant, constructed at a cost of over $5 billion, is to be torn down without having generated a watt of electricity. This action is being taken by the government of the state of New York in opposition to the federal government which, through its Nuclear Regulatory Commission, had given the plant

1

permission to generate the much-needed electric power on Long Island. Regardless of the final disposition of the nuclear plant, this incident illustrates a central theme explored in some depth in this book: what is risk, how can it be measured, and how can we know whether to build—much less operate—a project such as Shoreham?

This is not, however, another book about nuclear power plants. The arguments herein are much more general, exploring as they do the *fundamental* issues that surround such policies. In order to begin to appreciate what policy ought to be, one has to understand that the twin modern techniques we currently rely upon to guide our decisions in such matters are both completely incapable of generating the certain answers we desire. These two techniques are generally called risk assessment and cost-benefit analysis.

In theory, these procedures allow us to make decisions from among a set of policies, eliminating both those that are unacceptably risky and those that do not generate net benefits over the life of the project. This entire process rests on the objective tradition in science, uses a set of epistemological beliefs taken as true, and (ultimately) flirts with the ancient fact-value dichotomy in philosophy. Those who carry out these analyses, like those who write and teach statistics, know full well—and may even enlighten you should you care enough to ask—that the inductive techniques of science cannot provide any definitive answer for our doubts about whether some policy *ought* to be undertaken. But then, just like our statisticians who remind us that correlation is not necessarily *causation* but then act as if it is, those who carry out these statistical techniques turn to their analyses and their results to argue for or against particular policies. The statisticians (social scientists), meanwhile, argue that their latest regressions "prove" some particular relationship while also "explaining" some amount of variation, all the while forgetting that statistics can do no such thing. Statistics and statisticians can, in fact, prove absolutely nothing in this way.[4] Americans have been told otherwise so often that it is by no means uncommon for entire conversations to turn on nothing but a bunch of numbers, most of dubious accuracy, hurled furiously by each proponent at his equally bemused opponent. This is what passes for intelligent discussion as we close out the twentieth century. We have turned our public and private discourses over to expert scientist-advocates and their constituencies and allowed them to design, evaluate, and implement our policies. In all honesty, do you believe that our institutions, their operation, and their value have improved over your lifetime? If you find yourself answering "yes," I would venture to suggest that you are merely confirming my argument as something like the following thought courses through your mind: "What *evidence* does he have for these contentions?", or "On the preponderance of the evidence, I would venture an affirmative answer to his question." As I wrote above, we are now a nation addicted to inductive advocacy. It will be useful for readers to study the origin of this faith in objective, scientific technique as well as read a critique of that technique in terms that they can apprehend. This book is neither an attack on "science" nor on inductive procedures generally. It is an attempt to help readers to acquire a better understanding of the modern public policy environment: its philosophical roots, current technical apparatus, and actual (as opposed to predicted) outcomes. More specifically, it analyzes the divergences between policy promises and policy outcomes, and asks whether this gap in unfulfilled expectations can be narrowed.[5]

Another line of argument that will be pursued in this book concerns the politicization of science itself. Historically, any society that has accepted without question the policy decisions of its intellectual elites has invited their corruption. Techniques that stamp a seal of approval on projected policy will inevitably become corrupted over time as individuals and groups pursue their subjectively perceived economic advantage at one another's expense.[6] The fact that many of these individuals are scientists, natural or social, is hardly relevant. They are motivated by the same desires and biases as any other person, and they exhibit the same human weaknesses from pettiness and envy to desire for wealth, fame, and power (to say nothing of absolutely pure political motivation).

It is not, therefore, surprising that the expanded role played by scientists and policy analysts has been accompanied by great skepticism on the part of ordinary people regarding their pronouncements.[7] Indeed, if readers examine the print and broadcast media carefully, they will find in short order that two of the most ubiquitous words there are "study" and "critic." Everyone has a "study" that they say proves their contentions are true. Inevitably, there is a "critic" who has a different study that proves the opposite. The lawyer's threat, "you get your psychiatrists, and I'll get mine," has become *the primary method* by which we seek answers to policy questions—only the psychiatrists now must share the work with every other conceivable analyst-expert: from ecologists, "environmentalists," counselors, psychologists, criminologists, economists, sociologists, political "scientists," physicists, and biologists, to a host of inflated group titles such as policy analyst (usually prefaced by "senior" or "resident"), consumer activist, community "leader," resident "fellow," or media analyst. Quite simply, all this expertise has succeeded in accomplishing nothing so much as it has made it virtually impossible to tell the serious, knowledgeable individual from the army of bogus, politically motivated hucksters that dominate our modern system of information dissemination. Every person *must* have a title, which explains the absurd words accompanying the evening newscasts superimposed beneath the names of those individuals currently speaking (e.g., "neighbor," "consumer," or the recently coined "flag burner").

The justification of policy is a fact in every nation, be it democratic or totalitarian in form. In America, such justification now typically comes from allegedly disinterested studies provided by policy/science experts. We have erected vast monuments to their supposed powers: the National Science Foundation, the National Cancer Institute, the Centers for Disease Control, the Environmental Protection Agency, and on and on at both the federal and state levels. We christen our bureaucracies with the results we expect them to achieve, then move on to other matters.[8] What then remains? These large, well-funded machines construct a network that links researchers to taxpayer funds, and those to political and business constituencies and to the Congress itself in order to guarantee further funding.[9] It is looked to by the media as a source of enlightened opinion, and a source of exposé stories usually triggered either by some real-world disaster or some former employee "whistle-blowing" on his former employer. Inevitably after such episodes, even more funding and power is granted to the agency so that it can "carry out its mandate." Chapter Four of this book examines one such agency, the Centers for Disease Control, and its failed attempt to live up to its name in the case of the mythical Swine Flu pandemic of 1976.

Before reaching the final chapter where public policy meets reality, readers will have to absorb some theoretical material that will, the author hopes, make the Postscript much more understandable. Chapter One begins the analysis of the objective/subjective schism as it affects risk assessment today, demonstrating that the foundation of that endeavor, Classical-frequentist probability theory, is insufficient for the task to which it has been so often applied. Chapter Two explores the roots of the objective/subjective debate in the late nineteenth century between British economic theory, German historicist dissent, and the birth of the Austrian school. This overview will prove useful as the reader tackles Chapter Three and its discussion and criticism of cost-benefit analysis, a technique founded on the Neoclassical (British) tradition in economic theory. The critique that is offered is explicitly Austrian and is therefore subjectivist. All this will, in turn, come together for readers in Chapter Four; it is hoped that by then they will understand why today's policy debates are so often less fruitful than anticipated because of the true nature of the alleged power of modern, inductivist-driven scientific social policy machinery.

Over the course of this book, readers will also come into contact with some of the continuing disputes within economics, philosophy, and science in general. Keep an open mind, and realize that there may be more than one path to valuable, true knowledge of the world and its inhabitants. The scientific rationalism that has dominated the world in various manifestations since the Enlightenment is powerful and useful for some problems, but lacks persuasive power when applied to others; no matter how cleverly, honestly, or rigorously it is carried out, it cannot free us from other decision criteria.

An example may be helpful here. One of the turning points in the 1988 presidential election happened when George Bush pushed Michael Dukakis hard on whether the governor would support having children in public school say the Pledge of Allegiance each morning. Dukakis's response, that "any first-year law student knows" that mandatory recitations of the Pledge of Allegiance are "unconstitutional," struck many voters as an unsatisfactory answer. Why? The first thing to notice is that Dukakis dodged the issue by appealing to expert authority: in this case, the Supreme Court. That he was incorrect in his ascribing to the Court a position that it has not, thus far, enunciated is not germane. The point of the illustration is to show that, no matter how far back we regress our answer, eventually *someone* has to decide the issue on purely *normative* grounds. Increasingly and (I would argue) unfortunately, people are approaching every issue from the perspective Dukakis used. There is no "right" and "wrong," no "good" and "bad." There is only expert opinion that sets the parameters in our public and private lives, that allows us to burn the flag but not the leaves from our yards, to drive to work but not to smoke once there, to consume dangerous prescribed drugs but not apples with Alar, to drive under the "influence" of chemotherapy but not alcohol. Daily, we accept each new regulation because we think that policy experts have solved the issue of "do we or don't we?" once and for all through the application of scientific procedure. *They have not,* and this book will attempt to explain why this is true.

The theories, disputes, and critiques should all come together for readers in the analysis of the swine flu decision in Chapter Four. Here was, theoretically, a perfect opportunity for the "government as scientific policeman" model that has been proselytized so strongly and for so long to prove its validity; the arguments presented here will allow the reader to decide whether such a model should be retained. The analysis that follows represents both an introduction to and overview of deep problems in this view. This book is not a technical guide for readers to learn how to carry out risk assessment, nor is it a cost-benefit text. Readers can decide for themselves about the validity of the regulatory approach after having read the arguments presented in this book.

1

Risk Assessment, Probability, and Fact: An Overview

My thesis, paradoxically and a little provocatively, but nonetheless genuinely, is simply this: Probability does not exist.

B. de Finetti
Theory of Probability

One could safely claim that the need to make major societal decisions by explicitly using probabilities is a unique feature of our time.

P. M. Haas
Accident Analysis

This chapter will examine a technique that has come to be known as risk assessment. Risk assessments play an increasingly important role in the life of the average person, since they tend to determine the regulatory decisions that are ostensibly made in order to protect people from risks that occur in the (external and internal) environments that they live in. As ever more dangerous technologies are developed and implemented both by advanced and underdeveloped nations, there has been a concomitant growth in the use of risk-assessment methods. This increase in the application of risk-assessment technique has generated significant refinements in methodology and utilization of risk data. Nonetheless, vast areas of uncertainty remain unexplored by current approaches, and it is here that subjectivity usually enters the methods and arguments of risk analysts.

A complete understanding of risk assessment techniques is vitally important not only for policy analysts but for the framers of public policy as well. It is not that analysts are misusing existing methods, although that can be a problem. Rather, it is a lack of understanding of the broader philosophic questions that underlie these techniques which often leads to false claims concerning their efficacy, and therefore to invalid predictions and the loss of the public's confidence in the analysts' methods and/or alleged scientific detachment. This is not merely an esoteric issue confined to the academic community but an exoteric issue of considerable import. The expenditure of billions of dollars can depend upon public understanding and acceptance of risk assessments.[1] Part of the mistrust of scientists by the average citizen has been caused by the claims of scientists to know more than they actually do, as well as their inflated prognostications concerning the risks involved in various endeavors.[2] An understanding of current risk assessment technique can lead, therefore, to a better understanding of its limitations and proper application.

Comparative Risk Assessment

Comparative risk assessment (hereafter CRA) is a method used to make informed judgments concerning both policy choice and policy implementation. If information were perfect, every risk could be calculated with absolute precision. Yet even so, the need for CRA would remain—because judgments as to alternative policies require a conceptual framework from within which the anticipated effects of those policies can be both weighed and evaluated.[3]

All social policy involves trade-offs, not only between various risks but also between alternative resources uses and their attendant income redistributive effects.[4] All public undertakings harm some and help others. CRA is simply one more tool in a collection of techniques that seek to answer the fundamental, though elusive, question: is this proposed policy in the interest of the public?

The methodology of CRA can take any of the following forms:

* Engineering safety assessments
* Exposure assessments
* Health effects assessments
* Environmental risk assessments
* Economic and social risk assessments

An engineering safety assessment (ESA) is an attempt to determine the dangers "inherent" in particular physical circumstances.[5] An exposure assessment (EA) is an attempt to isolate the danger(s) to humans (and sometimes to animals) caused by coming into contact with certain things. A health effects assessment (HEA) seeks to quantify the expected negative health effects of such exposures as those found by an EA. An environmental risk assessment (ERA) seeks to define what negative (positive) effects can be expected to occur from the implementation of a particular policy. Finally, an economic and social risk assessment (ESRA) seeks to predict the effects, economically and socially, on individuals from the implementation of a particular policy.[6]

The methodologies employed in each CRA are a function of what is being attempted within each division of study. For an engineering safety assessment, there are four basic methodologies: (a) experimental testing, (b) mathematical modeling, (c) systematic approaches, and (d) engineering judgments. Experimental tests can be viewed within this framework as an attempt to learn of such things as "failure rates" for various circumstances. This is done by using samples under predicted conditions of use and generating frequency data for extrapolation to the universe for which the study is being conducted.[7] Mathematical modeling can be used, usually with the aid of computer generations, where sufficient data for individual circumstances are known or where the systematic relationships in various "systems" are well-documented and predictable. These models can then be used, at least in theory, to set such things as "safety parameters." Systematic approaches are typically based on decision tree analysis, a procedure that attempts to follow the possible consequences of individual paths throughout an entire system.[8,9] Engineering judgments are attempts to replace objective knowledge, which may not exist and cannot be generated except by expending large amounts of time and money, with the subjective judgments of alleged experts.[10] These four procedures, taken together, attempt to give the risk analyst an accurate picture of the "risk" of a particular methodology or system. They might be applied to a question such as: what is the likelihood that, in a standard water-cooled nuclear reactor, water will fail to be supplied to the central fuel core?[11]

Exposure assessments tend to be multidisciplinary undertakings where data from many sources is refined and then analyzed.[12] Among the most common sources are: (a) production and distribution patterns from industry, (b) occupational and residential lifestyle data from insurance companies, (c) data from government agencies, (d) data provided by academic research, (e) public health agencies, and (f) mathematical models of exposure and spread (or effects) that can be generated from actual sample data using computers.[13]

Health effects assessments require actuarial data, controlled experimental data, or epidemiological data on humans and/or animals.[14] Similarly, health assessments data and computer modeling can be combined to make guesses about environmental effects.[15] Environmental risk assessments utilize existing ecological models, data, and theory to project the expected impact on the external environment of contemplated policy changes, where such changes will or might have spillover effects.

When estimating the possible effects of a policy, economic and social risk assessments consider such sources as (a) financial planning models,

(b) past corporate risk-taking, (c) portfolio management and loan information, and (d) models of individual decision-making and welfare.[16] This type of undertaking is multidisciplinary in theory and practice:[17]

> Social risk assessment makes use of sociology, political science, economics, psychology, public health, education, religion and philosophy.

This is a "quality of life" multidisciplinary assessment and, as such, is necessarily less quantifiably precise than the standards for which those techniques previously mentioned rigorously strive. It is at this point that cost-benefit analysis may enter the risk assessment picture, in addition to such theoretical modeling techniques as game theory, operations research, systems analysis, utility theory, cross-effect impact analysis, technology assessment, policy analysis, judicial interventions, and public citizen input.[18]

This entire process of policy evaluation can best be summarized by cataloging the major institutional inputs as they relate to the mandating of CRA:

* Executive orders at the federal and state levels
* Legislative enactments at all governmental levels
* Judicial interventions at all levels
* Media input that does/does not convey accurate risk data
* Voter input through political impediments, election and recall of judges/legislators, and referenda
* Public attitudes towards "risky" policies
* Preconceived inclinations (biases) of those funding the study/evaluation

Ideally, a CRA should perform certain vitally important tasks: (a) compare the proposed technology and its risks with alternative technology, or with the *status quo ante;* (b) compare possible risk reduction with the reduction(s) made possible by alternative technologies; (c) compare the risks and costs of the policy under consideration with those of other alternatives as well as stating the benefits of the proposed technology with the risks, costs, and benefits of alternatives; (d) beneficially influence the decision process itself; and (e) show the natural risks involved in the *status quo ante* as compared to those reductions and increases in benefits that might flow from implementing the policy under consideration.[19]

A CRA is not carried out *a priori* and divorced from secular considerations. It needs to be understood that all CRA's are performed within an existing social environment that includes standing laws, executive orders, judicial interpretations, international treaties, and public attitudes.[20] Any cost-benefit study should attempt to reflect, as accurately as possible, the following: who exactly will benefit, and to what extent? Who will bear the cost, and how much cost? In sum, whose net positions are to be favored?[21]

Many critics have attacked CRA and cost-benefit analysis by appealing to the actual or alleged bias of those carrying out the calculations.[22] This line of criticism is unfortunately too often on the mark. The critical analysis that I shall develop will not focus on analysts' motivations, but rather on the current techniques that analysts use. Even assuming perfect intentions and no bias on the part of analysts, it is a contention of this examination that the *techniques themselves* still produce inaccurate results.

Comparative Risk Assessment: Problems and Possibilities

The first problem encountered when attempting CRA is the "data." The final output of a CRA is most vulnerable to criticism and questioning when data are either sparse or nonexistent. One reason is that individuals fear the "unknown." Subjective perceptions of danger, and of exactly how dangerous a thing can become, partially depend on people's suspicions of the analyst's claims to certainty. Yet frequently the analysts only have highly speculative techniques based upon myriad assumptions about situations that, in some cases, have never existed.[23] Another factor that tends to complicate calculations and create disagreement over the projections is whether or not "risk" is an inherent attribute of concrete characteristics or a set of subjective perceptions about inter-relationships between things and people.[24] No thing possesses objective risk; its "riskiness" is, rather, generally a function of how it can be used by human beings—hence the ever-present subjectivity in all risk assessments.

An example of the confounding effects that analysts cannot always correctly anticipate is traffic control procedures—specifically, the counter-intuitive outcomes that sometimes occur when the most obvious course of action is undertaken in order to make something safer. For example, the obvious policy to follow if there is an intersection that generates a disproportionate number of accidents is to install traffic lights. After the lights are installed, however, it sometimes actually *increases* observed accidents.[25] Outcomes such as these have led analysts into some rather speculative model building in cases where human interactions can affect possible outcomes. These various models of predictive outcomes can be classified and will typically fall into one of the following categories: (1) engineering models within which risk is objective and predicted outcomes involve only alleged relationships between non-human entities, (2) economic models that predict a trade-off between one possible outcome, "improved safety," and alternative outcomes whose occurrence might otherwise seem counterintuitive, and (3) risk homeostasis models that predict redistributed, rather than ameliorated, risk outcomes.[26] All these approaches can be useful in attempting to predict and/or understand the reasons why data often seem to tell peculiar stories. For example, consider the case of crosswalks on streets as attempts to reduce pedestrian-vehicular contacts. Such contacts are six times more likely to occur in crosswalks and, even when adjusted for the fact that more crossings occur at crosswalks, twice as many of these accidents occur within crosswalks than do outside of them.[27]

There is always a subjective element present in CRA, regardless of whether data are good or bad, or whether the question at hand is "simple" or not. Even where relationships are rather well-defined, as in toxicology, uncertainty remains. Dose/response data are averages, and individuals respond differently because of their not easily understood internal mechanisms, which in turn rely on the interaction of both the body and the mind.[28]

Subjectivism and Objectivism in Probability

The second problem any CRA must deal with is the uncertainty that permeates virtually all aspects of human existence. Basically, the estimation of risks can be accomplished in one of two ways: deductively or inductively. Although many people have come to associate scientific undertakings with induction alone, there is always a good deal of deduction present in standard scientific technique. One can estimate risks deductively by using the currently standard approach of Classical frequentist (or Laplacean) probability theory. All deductive (theoretical) systems that are completely specified—e.g., odds in a particular poker game—can be used as proxy odds for actual outcomes. The success or failure of this approach in risk assessment will be addressed below.

Alternatively, one can use an inductive approach in which data are collected and risks are measured from actual outcomes. This general way of attempting to carry out risk assessments is usually called either Bayesian statistics or subjective probability theory. The first approach calculates frequencies of predicted occurrences *a priori;* actual events are then anticipated to parallel the projections closely. The failures of this approach, which were taken to occur when unexpected events happened, led to the attempt to incorporate *actual* as opposed to *theoretical* distributions by collecting frequency data from the real world.

Were information perfect, then only one approach to probability would be necessary: the Classical frequentist method.[29] This method, like all closed mathematical systems, is true by definition. Classical theory cannot tell the analyst what the number will be when a die is tossed, but it can—in theory—specify all the possible outcomes in such a way as to "predict" that each side has a one-in-six chance of occurring.[30] Were all questions involving risk amenable to this approach, the analyst's task would be a good deal easier to accomplish. Unfortunately, the physical world can take a nasty turn now and again; it can leave theories "refuted" and even forgotten altogether.[31]

The first point to remember when examining the controversies in statistical analysis is that probability theory is not "science." Operationally, it is not different from Euclidean geometry: both have many useful applications in the actual world, but neither are inductively generated or empirically "falsifiable."[32] The truths of Classical probability theory are no less true than the Pythagorean Theorem, but they cannot be turned into "facts" by collection of data.[33]

In general, probability is divided into two broadly defined approaches: the objective and the subjective. The objective approach is associated with the Classical frequentists, the subjective with the so-called Bayesians—the name being derived from Bayes's Theorem.[34] (Bayesians themselves subdivide into objective and subjective camps, but that would carry the discussion too far afield.[35]) The primary difference between the Classical and Bayesian schools rests on the issue of what information is most important when a probability calculation is carried out: the purely theoretical, or the theoretical mixed with inductive knowledge?[36]

The Bayesian insight is used in risk assessment in order to deflect two possible criticisms that can be raised against the results of the risk assessment technique.

1. Data are incomplete.
2. Risk is perceived subjectively; it is not objectively intrinsic to objects.

To some degree, Bayesian analysis has helped to soften the first criticism; however, it has not really dealt with the second, because Bayesian techniques also assume "facts not in evidence." When data are complete, Bayesian and Classical methods both yield the same outcome. Everywhere else, however, the final number(s) will diverge.

To understand what is at issue, it will be necessary to examine the approaches of both techniques, so that a judgment can be made as to which method is best for the operationalizing of actual risk assessments.

Classical (frequentist) theory requires both *a priori* assumptions of certainty and non-deviational outcomes in the physical world.[37] As such, its operational axiom can be stated as follows: the singular outcomes (successes) divided by the total possible outcomes equals the probability of an event's occurrence, assuming that the event is "random" and the process is "fair."[38]

Consider the following example:[39] we wish to know the number of new tanks that an enemy nation intends to produce. A spy gives us information that the tanks will be numbered from 0001 up, and that production will total 1,000 units and then be abandoned. However, we think that his transmission might have been garbled when the number was given; for this reason, we suspect that the spy might have wanted to communicate that the production would really end at 10,000 units. Shortly after the transmission, we capture a tank and find that the serial number on it is 0998. Which transmission did the spy actually send?

A Laplacean concludes that the observed value lies within the assumed range of 0001 to 1,000, and assumes that this represents strong evidence for a maximum production of 1,000 units. A Bayesian might view this as a sample of one from an unknown universe. While the Laplacean has to carry the excess baggage of prior expectations which will color the interpretation of the observation (here, the serial nmber we have), the Bayesian is under no such constraint. He will form questions like the following. How likely is it that a tank with this serial number would have been captured if the run was actually 1,000? In light of this sample information, ought we to view this tank as evidence for the 10,000-run theory? What is actually happening here is that the Bayesian analysis relieves us of our prior biases that went along with alleged perfect information. It allows us to reformulate our calculations as we move along and acquire more knowledge. This process can be summarized as:

prior information and new information = revised estimates

Whereas the Classical frequentist analysis can only accept or reject a proposition based upon "sufficient experimentation," the Bayesian approach allows for greater flexibility in the light of actual outcomes.[40]

Contained within the Bayesian approach is information that can be termed "subjective" in nature. That is, Bayesian analysis does not arbitrarily exclude human intuitive perceptions from its operational base. Furthermore, this new information, while it can "confirm" or "fail to confirm" the objective analysis, is relevant to Bayesians not for that reason but because it is seen as new, relevant information about actual probabilities in the physical (as opposed to the theoretical) world.[41] It is for this reason that Bayesian techniques have begun to play a role in the estimation of very rare occurrences when an event (or, in classical terminology, "success") appears much sooner than the theory predicted it would.[42]

Despite this change in basic technique, Bayesian analysts must still rely on *a priori* statistical assumptions about the behavior of the physical world, including assumptions about a constancy in certain functions and about such constructs as the "large number law."[43] The Bayesians contend that

Bayes's methodology is the only statistical procedure that is consistent with true induction, since it processes information about the physical world and uses it to expand what is known about the universe, rather than proceeding from axiomatic assumptions that may not hold when tested.[44] Parry and Winter summarize:

> In Classical analysis the choice of action to be taken after a particular outcome has been observed can be evaluated only in relation to the requirements placed on a complete decision rule that stipulates what action should be taken for every outcome that might have been observed; for this reason, Classical analysis usually depends critically on a complete description of the experiment that produced the outcome because this same experiment might have produced some other outcome. *Induction, as learning from experience, is just the process of revising probability estimates in the light of additional information.*[45]

To make this issue concrete, assume that one is attempting to make a projection about an uncertain, unique event. Using the Classical (frequentist) approach, if the data are not perfect or reasonably complete, the result can be both wide of the actual estimate by a considerable margin and at odds with the subjective experiences of those who have prior knowledge of the subject matter of the analysis. If, after Classical theorists calculate that something has a one in one billion chance of occurring every one hundred years, it then occurs the following year, they are confronted with the option of either standing by the prediction by declaring that the event was indeed "random" or abandoning their prior analysis.[46]

One fundamental difference between Classicals and Bayesians seems to be that the Classicals see probability as *inhering in the things being analyzed,* while the Bayesians see it as states of subjectively held beliefs about those same objects and their relationship to humans.[47] This controversy has been argued in the literature of risk over the last two decades, yet it is still capable of arousing a great deal of emotional disagreement— especially now that risk assessments typically deal with unique and infrequent events. Bayesians have argued that their techniques are uniquely positioned to deal effectively with these types of assessments, and that Classical tools are less well-suited to the task. The Classicals continue to contend that the Bayesian approach not only is not superior, but may actually be inferior to the standard type of probability analysis.[48]

Basic Probability Issues in Risk Assessment Interpretation

Risk assessment cannot be operationalized without the use of the statistical technique called probability theory. Within the body of probability theory itself, as discussed above, there are disputes between various schools of thought, but that is of secondary importance for assessing riskiness. Primarily important is that some systemic method of probability analysis exist.[49]

There are two distinct methods by which one can proceed to "do probability," the theoretical approach and the empirical approach. In this regard, probability theory is really no different from other scientific endeavors that attempt to find and use true knowledge about the world. Because so much risk assessment is concerned with events that have not as yet occurred, or have occurred only in isolated instances, the two techniques are generally used together.

As with all pure theory, the objective *a priori* approach can be done with a minimum concern for empirical observation and/or empirical relevance. In this regard, it is helpful to think of it as simply a branch of applied logic or mathematical theory. Just as no one attempts to falsify Euclidean contentions about nonexistent entities (such as parallel lines) by actually bringing in "evidence" that they do not exist, no one really tries to falsify the contentions of theoretical probability analysis. Like many other theoretical systems, it is true by definition. The entire distribution of outcomes of die-tossing can be theoretically modeled without any actual die ever being tossed.[50] Suppose, however, that one were to model the actual tossing of a die, with a view to ascertaining whether it was "fair" or "unfair." In this case, certain problems necessarily arise, although a modification of the original deductive model that transforms it into a purely inductive one might give us a more accurate picture of the results of a long run of tosses of this particular die.

In the theoretical approach, all factors that can affect the outcomes will be "controlled for." The die will be, by assumption, "fair." That is to say, no one result can have a greater tendency to occur than any other result. The die will not be allowed to do certain things—e.g., explode in midair, land on its edge, disappear, or get lost. In this way, the entire analysis will be internally consistent and logically irrefutable.[51] No actual die tosses from the physical world can ever be used as data in order to accomplish a refutation of the model.[52] A further assumption (one usually left unstated) is that the entire "system" in which the tosses take place, either actually or in theory, must be disturbed as the process unfolds. The truth of this can be seen easily enough, for identical trials would produce identical results. At least one variable must be allowed between "tosses," though variation must be "controlled" so that the system will produce the expected outcomes.[53]

The advantage of this objective approach is that it provides humans with absolute certainty about the range of possible outcomes which, by their very nature, inductively generated models cannot secure. There is a trade-off here, however, and it is that the inductive approach might produce better results and, therefore, be more predictively useful than the theoretic approach when reality fails to conform to the assumptions of the theoretic approach.[54]

The important point here is that we can never confirm the propositions of classical probability theory, no matter how many times we toss a die, because the outcomes cannot be inductively used to refute the model. Perhaps (someone defending the model might say) we have simply not tossed the die enough, for no matter how "rare" the observed outcomes are the original model "predicted" them and attached to them a positive probability. If this argument sounds circular, it should; there is no escape from this logical box.

Suppose that some outcome of tossing a supposedly fair die was predicted to be very unlikely, say one in one million. Even so, no perfect case can be made that the die is not "fair," or that, in rhetorical terms, reality is not conforming to the theory's assumptions. To leap to such a conclusion would be to allow inductive outcomes to override theoretical specifications. How could we generate a stopping rule for these situations? In fact, to say that a die is "biased" has no meaning apart from the contention that it does not behave as the collection of theories assumes that it ought to be behaving. Yet these theories give us no sure way to know this, nor do they supply a way to stop short of infinite tosses.

There is another possible source of error inherent in applying analytic techniques to synthetic knowledge, and that is the ever-changing nature of reality. No matter how much data we have about some state of affairs, it may behave differently in the future. In the case of the die, it actually might be "biased" when we cannot say so with certainty, or it might not be when we claim that it is. All this should be seen immediately as variations on Type I and Type II error in statistical theory, where reduction of the probability of committing one type of error merely increases the chances of committing the other.

Bayesians argue that no matter how carefully we try to specify our theoretical models, some subjective, interpretive elements are bound to filter into our analyses. They can be termed "human intuition." This is true of all scientific undertakings, whether deductive or inductive in nature. Some probability theorists have claimed that, in fact, there is only subjective probability.[55] It remains helpful, however, to retain the objective-subjective distinction within this argument because it is consistent with the philosophical problems inherent in all scientific undertakings.[56]

The fact is that all evidence is filtered through human perceptual faculties, and this can lead to difficulty even when the "evidence" is in front of our eyes. A recent example is the famous case of Love Canal, New York.

Biomedics, a Houston-based firm, reported that residents at the infamous toxic waste site were suffering from "chromosome damage." This was widely reported in the media. In a study of Love Canal done under the auspices of the state of New York, several scientists attempted to verify Biomedics's conclusion. At first, Biomedics would not make its slides available for peer inspection. Later, after the panel had viewed the slides, it was reported (not so widely) that no evidence of chromosomal damage could be found on the slides beyond what was "normal." Further, some of the panel members did not even see chromosomes at all![57]

It would be tempting to dismiss this case as some kind of scientific "aberration," a particularly lurid example of bad research being uncovered by peer review, or dishonesty not paying off in the long run. But questions remain. Did the waste affect the chromosomal structures of the residents? Did Biomedics find credible evidence for such damage? Was the panel correct, or did they overlook evidence, or were they covering up in some way? The lack of a clear-cut outcome in this case is not a rarity; it is the norm in public policy disputes. The *New York Times* might editorialize that people were injured more by the false publicity surrounding Love Canal than they ever were by the wastes themselves, but who really believes this, and how could the *Times* know it with certainty?[58]

One can agree with those analysts who believe that there are "facts," that those "facts" can be known with certainty, and that the theoretical probability approach competently models such "facts." If so, one is led to conclude, along with these analysts, that "risk" is objective and hence numeric, as well as that its existence lies outside of human perceptions and wishes. It has to be admitted, however, that these hopes, wishes, and fears often can be a confounding element in any analysis of risk. This has happened often enough to suggest that it is a good deal more than the fantasy of subjectivist theorists.[59]

Throughout human history, scientific "fact" has proven to be false so often that it would be foolish to maintain that we can ever completely eliminate error. For precisely this reason, scientists are usually modest in their claims. Laymen are often skeptical of the claims of certainty that scientists make, partly because of the historical record and partly because they have strong, intuitive feelings about risk. Analysts can explain the theoretical approaches scientists take with two arguments, but neither give comfort to non-scientists. First, there is always the "imperfect information" argument. Some unforeseen, exogenous factor affected the assessment. Second, scientists might claim that prior estimates were "biased" by the predisposition of earlier analysts to reach some cherished conclusion. The first defense, while obvious, remains unsettling; after all, information is never perfect. The second defense is not flattering to scientists, yet it could be true more often than they would like to admit.[60] Special pleading cannot, of course, ever be completely eliminated from the policy process.

A third set of possible explanations for the tendency of the actual physical world to be imperfectly modeled by our theories is the interaction between the subjective, internal environment within human beings and the external environment over which they often have limited control. It is now well documented that people can make themselves sick, can help make themselves well, and can even kill themselves by passively wishing it to happen.[61] The possibility of these effects confounding our analyses suggests that risk analysts ought not to reject the subjective perceptions of affected human populations out of hand. The question for analysts, then, is *how* do we incorporate this into the technique(s)?

Return for a moment to the Love Canal episode. It is not inconceivable that the residents' health began to become adversely affected from the moment they learned of the buried chemicals. To the extent that this can be true in such instances, all objective risk assessments of their current and future health states will not be able to pick up these self-generated, negative, fear-driven outcomes. Rather than discounting these possibilities, and trying large and expensive educational campaigns that attempt to educate people (i.e., convince them that they are wrong in their assessments of the dangers of agent *x*), analysts might be better advised simply to take such things into their probability models *ex ante*. For this reason, adjustments to the "riskiness" of undertakings should be made whenever it can be shown that there is intense and widespread emotional opposition and fear of negative effects. That fear is capable of producing some of those effects; to the extent this is true, analysts will continue to get incorrect estimates of health risks.

Estimation of the Risk of Unique Events

Because the final chapter of this book focuses on an allegedly high-risk and low-probability case, it may be useful to examine the techniques used in order to carry such estimates out. For example, insurance companies can be profitable undertakings because of the accuracy of class probability estimates. Although they can predict the probability of death for a class of policyholders (usually within rather narrow and well-specified limits), they cannot predict the moment of death for any one particular policyholder. This is true because no probability can be attached to a unique event; that is, an event that belongs to a class where there is only one member and no prior ones.[62]

Given the nature of this situation, it is surprising to find that most risk assessment is done on a case basis, rather than a class basis. To take a relevant example, we might formulate the following question: What is the probability that nuclear power plant x will suffer a LOCA (loss of coolant accident) that results in a total "meltdown" of its fuel core? Since there had never been such an accident at a commercial reactor before Three Mile Island (and not even then, exactly), the only sensible manner of approaching this problem was to calculate a long string of independent and contingent objective probability functions. The answer was allegedly obtained from the so-called Rasmussen Report, alternatively titled the WASH-1400 Reactor Safety Study, carried out under the auspices of the Nuclear Regulatory Commission. Over 275,000 man-hours of effort went into producing this report, and the answer given to the American people was that such an occurrence was very close to impossible. In fact, the probability was not significantly different from zero.[63] How did the analysts construct this estimate? The state-of-the-art risk assessment techniques were fed into the most sophisticated computers then available, and the final result no doubt soothed the fears of many citizens.

The study calculated a long chain of independent Classical objective probability functions. Some of these, however, relied not only upon theoretical assumptions but also on inductive input. For example, to understand that a LOCA can deteriorate into a worse-case meltdown, we have to assume that backup generators, for whatever reason, have failed to supply the necessary water to the fuel core. What is the probability (and how is it to be established) that two emergency generators will fail to supply that coolant? The beginning of the answer is generated by inductive data from actual generator start-ups and failures. This is combined with the assumptions from Classical theory that the failure of two generators to start is a "random event" where the two failures are "independent" of each other. The probability, then, that both would fail to start is the multiplicative of the two separate probability calculations. By stringing together a chain of such independent outcomes, the Rasmussen Report authors were able to generate their estimate of the probability of an occurrence like a major LOCA meltdown.

The outcome of this study can be debated on theoretical and empirical grounds. Many objections can be raised about the manner of procedure adopted by the panel of risk analysts, and other objections have become significant since the information associated with the Three Mile Island accident has been analyzed. For one thing, the report failed to include human error of the sort found in the Three Mile Island operating team in its calculations. Virtually every action taken by operators made the incident worse than it otherwise might have been.[64] But, to be fair, how is the risk analyst to place probability estimates on a series of observations that have never been observed? That is the central issue for estimation of the probability of a unique event.

One possible way of approaching this issue of human error would be to attempt an inductive generation of random error distributions. But this is virtually impossible to model, because it is next to impossible to observe random human error; further, we would expect that human error would not be random to begin with. It is simpler to develop decision tree-like structures where choices are distinct and delimited than it is to accept notions of humans acting randomly (and, all too often, destructively) enough so that they almost suggest actual intention to sabotage particular processes. We believe we know, however, that this is not the case.

Operationalizing Subjectivity in Risk Analysis

Each calculation, Bayesian theorists argue, carries a subjective element that can be termed "degree of belief" after Polanyi (1962) and Keynes (1973). This belief is, quite simply, the confidence that the analyst has in the validity of the calculation. If the assumption is made that the analyst's uncertainties can be modeled by a normal distribution, then absolute certainty can be seen as the probability that a calculation is certain (that is, $p = 1$). At the other extreme, a complete lack of confidence in the calculational correctness would result in a degree of belief, p, equal to zero. Each calculational step in any risk assessment procedure can be taken through this belief system for subjective valuation and verification. A simple formula can be added to probability calculations to allow for uncertainty attaching either to the data or to the theory through which the data is manipulated.

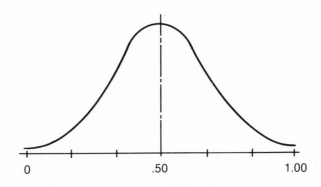

0   .50   1.00

Figure 2 - 1: Distribution of Subjective Uncertainty

Where it is argued that the probability is that a "good thing" will occur, then this approach would decrease that probability by an amount consistent with the subjective degree of belief of the analyst in the calculations. Alternatively, if the calculation is that a "bad" event will occur, then the estimate will be increased by a factor of one plus the analyst's perception of the certainty attaching to the calculation. To take an example, suppose an analyst is .75 certain[65] that the event will occur with a probability of .01. Then the probability of .01 is increased by 25% and the estimate is raised to .0125, a small but significant change nonetheless. In the case of extremely complex and, therefore, very uncertain outcomes, where the confidence cannot be more than, say, .10, the factor of multiplication would be significantly higher, in this case a 90% increase. This almost doubles the estimated rate of occurrence; it might well affect the way in which particular programs or outcomes are viewed. Increasing the probability of bad occurrences should decrease the all-too-common explanatory apologetics required of analysts when their seemingly certain predictions are refuted by real-world events.[66]

The application of this modification would lead to the accomplishment of two things. First, it would make the conclusions of risk assessors less prone to the charge that they are offering certainty where none is possible. Second, this approach might well improve the analyst's track record by increasing the accuracy of the predictions that the techniques are able to generate. It should not be a major problem to get responsible analysts to adopt the degrees of belief approach, even though some might abuse it by maintaining—as they now implicitly do—that each calculation somehow carries a p equal to 1. To the extent they forgo that type of scientific arrogance (and they actually do this explicitly many times, as when minimum exposure levels are set or bridges are built), the public will be better served as well as protected, and the status of the analyst's endeavors will increase proportionally.

2

The Objectivist-Subjectivist Debate and the Historical Development of the Austrian School

*Pure economics has a remarkable way of pulling rabbits out of hats—apparently **a priori** propositions which apparently refer to reality.*

Sir John Hicks
Value and Capital

All of the logical propositions of monetary theory would remain true even if no human being had ever used money.

Ludwig von Mises
Human Action

*The characteristic mark of Kantian philosophy is the claim that true **a priori** synthetic propositions exist.*

Hans Hoppe
Praxeology and Economic Science

The debate concerning the role subjectivity plays in natural science is also present in the social sciences. The essential argument is about what effects follow in our theories when subjectivity is explicitly incorporated into scientific models. This debate raged without resolution in the late 19th century; it continues unabated today as philosophers and historians of science demonstrate that current theory has often simply accepted objective methods without acknowledging that this ongoing debate remains unsettled. The modern Austrian approach to the study of economics developed along lines very different from those prevailing in the British Classical tradition. The major relevant difference, from the point of view adopted within this volume, is the Austrians' adherence to an explicitly

23

subjectivist methodology. Familiarity with the history and nature of this conflict is valuable because it enables us better to understand subjectivist criticisms of Neoclassical risk assessment and benefit-cost method. The dispute between the objective and subjective in science appears in both the natural and the social sciences, although the primary focus in this book will be on economic theory and its applications to public policy.

As was seen in the discussion of risk above, objectivists believe that reality is totally outside of human consciousness, although human reason can be a very accurate guide to that reality. Subjectivists argue that reality is not simply a collection of objects, standing apart from human consciousness, but a mixture of those objects and subjective perceptions of them. Whether one takes the objectivist or subjectivist position dramatically affects one's perceptions of risk. Objectivists theorize that risk is integral to the objects that are the focus of analysis. Conversely, subjectivists theorize that risk is the outcome of an interaction between human perceptions and the objects being examined.[1] To understand the nature of this variety of approaches to science, it is useful to examine the historical roots of the objectivist-subjectivist debate.

The Austrian school of economics dates from the publication of Carl Menger's "Grundsatze der Volkswirtschaftslehre' (*Principles of Economics*) in Austria in 1871.[2] The term "Austrian" did not come into general use until after publication of Menger's 1883 work "Untersuchungen uber die Methode der Sozialwissenschaften und der Politischen Oekonomie insbesondre" (*Problems in the Methods of Political Economy and Sociology*).[3] The great methodological conflict between the Austrians and the German Historical school had begun. The other two famous members of the Austrian school, Eugen von Boehm-Bawerk and Fredrich von Wieser, never studied with Menger, but they were drawn to his theories and methods through his writings. All three held chairs in Austrian universities by the time the so-called "Methodenstreit" (dispute over method) erupted.[4]

Menger was at the University of Vienna intermittently between 1872 and 1903. Von Wieser was also at the University of Vienna, when not at Charles University in Prague, between 1883 and 1923. From 1904 to 1914, his tenure at Vienna overlapped that of Boehm-Bawerk.

Boehm-Bawerk held the position of Finance Minister of Austria, and would receive much acclaim (and criticism) after publication of his monumental treatise *Capital and Interest*.[5] Von Wieser's major opus was *Social Value*.[6] He was, like his counterpart, a minister in the Austrian cabinet, holding the position of Minister of Commerce.

The focus of the conflict with the German Historical school was over the central teaching of Classical economic theory as it related to issues of special economic privilege enjoyed by certain individuals and interests in pre-capitalist society. It was a tenet of the Classical school that all such "artificial" barriers to the advancement of entrepreneurial activities ought to be abolished in the name of promoting the public interest by allowing individuals maximum latitude to pursue their private interests. Those incumbent beneficiaries of privileges thus attacked by liberal economic theory were, by predisposition and necessity, compelled to counter the British economic arguments. The German Historical school, that self-proclaimed "intellectual bodyguard of the Hohenzollerns," began an attack on British Classical social, economic, and political theory which continued for many years.[7] What could have caused such a dispute? Essentially, the question was whether or not economic theory dealt with actual experience or was merely idle theorizing. Since the historicists did not like the policy conclusions that seemed to spring forth from the Classical theories, they focused on the issue of method as a possible way of discrediting the British Classical economic policy prescriptions. The British Classical writings suggested a *laissez-faire* approach for government policy, and a seemingly ever-widening role for private market forces. This focus on the individual's happiness and autonomy, coupled with the narrowing of the state's proper role, did not please many of the German professors whose devotion to monarchy was proudly self-proclaimed.

Their first line of argument was that although Classical economics might be applicable in England, it was not applicable in Germany. The historicists emphatically denied that the theories put forward by the British Classical school might be universally true for all.[8] They argued relativistically that each situation was historically different, and that no general theory therefore would apply at all times and in all places. As Mises remarked:

> Thus economics in the second German Reich, as represented by the government-appointed university professors, degenerated into an unsystematic, poorly assorted collection of various types of knowledge borrowed from history, geography, technology, jurisprudence, and party politics, all larded with remarks about the errors in the "abstractions" of the Classical School.[9]

Beyond this, the English utilitarian tradition was not tolerated in German universities.[10]

The Austrians were also champions of free markets, arriving at that position from an entirely different methodological framework than than their British colleagues, yet just as anxious to proclaim the power and universality of the new science of economics. Ironically, although the British-Austrian assault on historicism was successful, the two uneasy allies would themselves begin a continuous dispute over method that continues to the present day.

German universities at the time of the dispute were totally controlled by the government, though owned and operated by various grand duchies. Professors were civil servants; as such, they had to obey the orders and regulations issued by their superiors, the bureaucrats in the Ministry of Public Instruction.[11] Interestingly, the important contributions in economics during this period—e.g., the work being done by von Thunen and Gossen—were done by individuals who did not hold university professorships.

In his second book, Menger explicitly rejected the epistemology of the German historicists. The historicists responded with a review penned by Schmoller. Menger's response came in 1884 in a pamphlet titled "Die Irrtumer des Historismus in der Deutschen Nationlokonomie"[12] (The Fallacies of Historicism in German Political Economy). These attacks and rebuttals came to be known as the "Methodenstreit."

The principal players in this debate were often confused as to what the essential issue was and how best to argue their respective positions. This is not at all unusual, especially when one can view the earlier exchange from the vantage point of the present, with its more sophisticated theoretical perspectives. According to Mises, the principal issue was whether or not a science was possible outside the discipline of history itself.[13] The prevailing *zeitgeist* in natural science at that time in Germany was "materialist determinism." The only path which allowed the possibility of augmenting human knowledge was experimentation in psychological and biological laboratories, since all human thought was taken to be a process of chemical reactions mixed with environmental stimuli, a view not that uncommon even in our own time.

Schmoller and his followers rejected materialist determinism *not* because they understood it well and could therefore out-argue its proponents, but because it contradicted the set religious tenets of the Prussian state.[14] The fact that they misunderstood the arguments of the materialists can readily be seen by their advocacy of a method remarkably similar to Comte's pure positivism. Even so, they disparaged the source of the method because Comte was a Frenchman and an atheist. In fact, a consistent application of Comte's position was not incompatible with materialist determinism in any form.[15]

World War I effectively put an end to the "Methodenstreit," while adding the triumph of the British Classical tradition in economics. That school had by then incorporated certain important developments in France, England, and Austria in the 1870s into its corpus. What finally emerged from these developments was the Neoclassical paradigm, which proceeded to incorporate many Austrian insights (especially theories of capital), even as it rejected Austrian subjectivist methodology as generally inappropriate for economic science.

Coincidentally, because of the rise of Nazism, both von Mises and his most famous pupil, F. A. Hayek, departed Austria. Mises came directly to the United States, while Hayek, who ultimately was to teach at the University of Chicago, stopped first in England to teach at the London School of Economics.[16]

It is interesting that Hayek's ideas on trade cycles, with their distinctly Austrian flavor, were just beginning to find an audience in England because of his presence there when Keynes wrote *The General Theory of Employment, Interest, and Money*. Hayek's views, which had aroused such interest in London in the early thirties and influenced such young scholars as Haberler, Robbins, and Morgenstern, were quickly forgotten during the halcyon days of the "new economics" or "Keynesian revolution." The fact that Hayek's trade cycle theories were supplanted by those of the Keynesians was to have profound implications for the Western democracies many years later.[17] If Hayek's analysis had been more widely examined and its implications sufficiently understood, policy might have been different in the 1960-80 period in the West. Serious inflations, with their attendant macro-dislocations, might have been avoided.

The Austrian school, because of the eventual residencies in the United States of Mises and Hayek, today calls America home. Mises's influence was felt through his prolific and powerful writings, as well as an informal seminar that he conducted for many years at New York University, though he held no official post there. Hayek was to influence many young scholars by their association with his Committee on Social Thought at the University of Chicago during the 1960s. There are now Austrian programs at both New York University and George Mason University, and there has been an increase in the number of younger economists who call themselves Austrians.[18]

Although the United States today has more Austrian economists than any other nation and more economists sympathetic to Austrianism as well,[19] their number relative to other schools still places Austrian thought in a minority position within the economics profession generally. Although firmly a part of the Neoclassical tradition, and defenders of that tradition against the historicists, Austrianism still is not generally well-regarded within the mainstream Neoclassical paradigm. The reasons are not hard to find. First, the Austrian subjectivist tradition has fallen out of favor as the positivistic methods of natural science have gained more adherents in the social sciences. Second, the classical liberal political viewpoint that Austrians generally defend has fallen out of favor in this age of large totalitarian nations aligned against massive welfare states. Between these two suggested reasons it is (surprisingly enough) the methodological dispute that has most divided Austrians and Neoclassicals. Neoclassical theorists often view Austrian ideas as some kind of "assault on science," as Austrians themselves once viewed the writings of the historicists. As White remarks:

> Radical subjectivism has been, in short, the distinctive method of the Austrian economists.[20]

Radical subjectivism is often seen, however, as a step away from objectivity and towards some non-scientific methodology in economics. Yet Hayek has maintained that every major advance in the history of economic theory sprang from a new application of subjectivism.[21]

Most economists, if they have heard anything of the Austrian method at all, believe that it is an old school whose primary claim to distinction was that its "father," Carl Menger, was one of the codiscoverers (along with Walras and Jevons) of marginal utility theory.[22] This discovery led to much exciting new work in economics as Neoclassical theorists attempted to import geometric/mathematical methods into economics from physics— since "marginal" (to them, at least) meant "small increments," and that meant, in the methods of the time, the application of the calculus.

Although Menger's first work was a pathbreaking effort—Wicksell, for example, remarked that no other work since Ricardo's *Principles* affected economists as much as the *Grundsatz*[23]—his works remained largely unexamined outside of Austria. Only the Swedish economists of that period seemed familiar with the Austrian tradition; this can be seen by the overlap in key theoretical areas.

Menger's method was, quite simply, "essentialism." His work was firmly grounded in Aristotelian metaphysics, Aristotle being quoted at great length in his second book—which was, after all, a reply to the historicists who had been critics of his first work. For Menger, individuals are driven by psychological needs that are independent of the reasoning process, yet must be discovered by each person through his own development. Choice arises due to scarcity and imperfect human decision-making, thus the subjective nature of the analysis. This quest for the essential, unchanging properties of the economy would infuse the thought and writings of all Austrians with Menger's basic methodology. As Lawrence White comments:

> At every step, Menger emphasizes and re-emphasizes the subjective nature of these properties, their dependence on the knowledge and attitude of the individual concerning his wants and the ability of objects to satisfy them.[24]

The Mengerian approach included an explicit and total rejection of mathematical methods in economics. Menger was fully trained in mathematics, but chose explicitly to reject its use in his writings because, as he wrote to Walras:

> How can we attain to the knowledge of this latter (the nature of value, rent, profit, the division of labor, bimetalism etc;) by mathematical methods?[25]

The search for essence in causal relationships precludes the possibility of modeling equilibrium systems which are by design mutually deterministic. Menger built up the theory of the modern economy deductively, from a one-person world to a complex, multi-person system. He expressly repudiated the idea of economic "equilibrium," the construct that formed the base of Walras's system, although he did not deny that the economy could move towards equilibrium positions.[26]

Menger's theory of consumer valuation was based upon a rigid adherence to the Law of Diminishing Marginal Utility. Each unit of a good was allocated to that use for which the individual felt the strongest (most urgent) need. Later, Neoclassicals would denigrate this insight by derisively calling it "psychological economics." Yet the insight remains. The satisfaction of subjectively felt need was done purely on ordinal scales through subjective choice. Value had no objective meaning except within individual minds, and values could not be added or subtracted, but only ordered. Although Menger uses the word "utility," he does not subject it to the types of procedures that ultimately led the Neoclassicals to manipulate utility functions as though they contained some objective thing that individuals actually maximized. The confusions caused by overplaying the analogy between utility maximization and maximizations in other mathematical endeavors eventually led economists to move away from the Mengerian construction of utility as *always being* marginal and *not* tied in some way to increments of goods. It further led them to embrace the idea that if marginal utility was the utility of a marginal consumption however defined, then it had to be (at least in theory) measurable.[27]

Stigler summarizes other differences between Menger and the British/ French codiscoverers of marginalism in the following passages:

> Each [Menger and Jevons] was, in contrast with Walras, essentially non-mathematical in method; each wrote on certain parts of economic theory but intended eventually to write a comprehensive treatise which never appeared; each was in sharp revolt against the classical political economy. But Menger's theory was greatly superior to that of Jevons: It was systematic and profound; it avoided the clumsy and unnecessary use of mathematics; and in particular, it generalized value theory to include a sound general theory of distribution.[28]

Stigler also notes that Menger made original contributions not shared by Walras or Jevons, including the principle of variable proportions and insights into the subjective imputations of value made possible by humans acting to achieve personal ends with observed means. Stigler is most generous in his summary on Menger's place in the history of economics:

> Certainly the most antagonistic cannot deny Menger a prominent place in the hall of economic fame, and the more enthusiastic, of whom the writer is one, will feel little hesitancy in acclaiming the *Grundsatze* as a treatise which is in fundamental respects unexcelled by any other between the *Wealth of Nations* and Marshall's *Principles.*[29]

The first systematic codification of the methodology of the Austrians was put forth by Mises in the introduction to his book *Human Action.* In a later work titled *The Ultimate Foundation of Economic Science,* he returned to these themes in great detail.[30] Interestingly, Mises ended up repudiating much of what Boehm-Bawerk had attempted to teach him. Unlike Boehm-Bawerk, Mises was very intersted in the methodology of science.[31] Mises began by developing the outlines of "praxeology" or, as he defined it, the "science of human action." A neo-Kantian, he expressly rejected the methods of induction as being incapable of generating truths about the world. His system of reasoning is almost entirely *a priori* in nature, and deductivist in method.[32] As White writes:

> Ludwig von Mises and Richard Strigl, retaining the ontological nature of Austrian theory, but placing it on a new epistemological foundation, led the formalist branch in deriving the subjective valuations of individuals only from their actual choices. To Menger's concern with needs and Boehm-Bawerk's with psychology, Mises objected that economics as a positive science is not concerned with the motives behind human actions but only with the implications of the actions themselves.[33]

Mises needed to bridge the gap between his deductivist, axiomatic approach to science and the actual, physical world. He attempted to do this by asserting that the logical structure of human action is directly linked to the logical structure of human thought processes.[34] But what of an original starting position, the original axiom(s)? Mises is quite explicit on this point:

> The starting point of praxeological thinking is not arbitrarily chosen axioms, but a self-evident proposition, fully, clearly and necessarily present in every human mind...the cognition of the fact that there is such a thing as consciously aiming at ends.[35]

Creative and insightful extensions of Austrian methods have been advanced by Nobel laureate F. A. Hayek. His life has been spent as much on methodological issues as on economic ones. He devoted an entire work,

The Counterrevolution of Science, to sustained examination and deep criticism of the techniques of the pure sciences when "misapplied" to the social realm.[37] Hayek faults standard economic procedures for their tendency by the use of such postulates as perfect information, continuous divisibility of goods and services, mutual determination of relationships, and assorted other "simplifiers" to assume away precisely the important and interesting questions that a true science of economics should be concerned with explaining. Specifically, this would be the coordination of vast markets and the complex set of interrelationships and foundational structures necessary for that coordination to take place under conditions of uncertainty.[38] The objective facts in social science are, for Hayek, completely theory-laden in a Kuhnian sense,[39] and entirely subjectively perceived.[40] These "facts" cannot be manipulated as facts are in the natural sciences; to attempt this will result, he claims, in only a "pretense of knowledge."[41]

The debt that Austrian economic theory owes to Max Weber has been explicitly noted by Ludwig Lachmann. Weber sought to hold up a different model for social science by his use of the method of *verstehen* (subjective empathy and understanding). According to Lachmann, Weber attempted this by striving

> to uphold the methodological independence of the theoretical social sciences from the natural sciences by stressing the cardinal importance of *means* and *ends* as fundamental categories of human activity.[42]

Even though the *verstehen* concept was first introduced as a method of history, Lachmann argues that the Austrian school, whether consciously or not, has been using *verstehen* as a theoretical method. For Lachmann, the task of economic reasoning is purely heuristic; that is, theory should "make the world around us intelligible in terms of human actions and the pursuit of plans." He writes:

> The task of the economist is not merely, as in equilibrium theory, to examine the logical consistency of various modes of action, but to make human action intelligible as all true economics is not "functional" but "causal-generic."[43]

The Austrian methods, then, are an attempt to understand economic "reality" through qualitative methods. The eschewal of mathematics precludes the use of quantifications and quantified predictions. This position is explicitly chosen by Austrians and is not a product of their mathematical igorance, a charge sometimes propagated by the Neoclassical detractor.[44] Austrians do not work on general equilibrium models because the focus of their technique is to understand change, but they refuse to make quantified predictions of changes because of their belief that the future is always "uncertain." Neo-Austrians, such as G.L.S. Shackle, have made very real contributions to the literature on uncertainty in economic theory, but the Neoclassical school seems transfixed by T. W. Hutchinson's lament:

> To allow expections and uncertainty into economic theory is to open a Pandora's Box of unsolvable problems.[45]

A further aspect of Austrian technique is that it is thoroughly "methodologically individualist" in nature. For Austrians, therefore, macroeconomics simply has no meaning, as its variables do not act causally on each other, nor are the theories firmly grounded in the assumptions of microeconomic choice theory.[46] Hence, they are unlikely to be employed with forecasting firms, or in the area of estimation of empirical demand, or within government agencies that endeavor to gather macroeconomic statistics. They reject the additive properties of economic variables, especially including (but not limited to) utility, costs, benefits, interest rates, capital, welfare...the list is virtually limitless.[47] Indeed, they argue that it makes no sense to speak of macroeconomic "variables" in a world without economic *constants*. In all this, their preference is for the study of individuals, not the attempted construction and measurement of collectivities. In this loyalty to the sound foundation provided by the assumptions of microeconomic choice theory and, therefore, to methodological individualism, they share in the criticism that has been directed at these macroeconomic techniques by anti-Neoclassical theorists.[48]

Today, the Austrian school has more adherents and sympathizers than at probably any other time in its history. Undoubtedly, this is due to the dissatisfaction of younger economists with the as-yet-unfulfilled Neoclassical promises such as "fine-tuning" and the outrageous claims that used to be made on behalf of positivist methods in economics.[49] Further disaffection can be seen in the massive failures of the centrally-planned utopian collectivist experiments that used to drain so much of the energies of the young, and the attendant failure of the large, mathematical macroeconomic models to predict accurately the course of events in the freer economies of the world.[50] All this combined with the award of the Nobel Prize to Hayek has, once again, boosted the fortunes of Austrians and generated interest in their ideas. What the Austrians will be able to do now to retain that interest and build upon it is a prediction that no Austrian, by method, would ever willingly make. It is clear, however, that Austrian theory is not stagnating.

Subjectivism: Scope, Limitations, and Current Status

Because science is performed by humans and flows from processes within human consciousness, there can be no limitation on the application of subjectivist methods. In fact, a consistent and radical subjectivism, aggressively pursued, can degenerate into the nihilistic claim that knowledge of reality is always transient. Recent arguments by some Austrians have others fearing that subjectivist methods have gone too far.[51] The great questions posed by philosophy are: how do we know anything, and how do we know that we know it? The subsidiary question—how do we choose among knowledge claims?—has bedeviled philosophers since the first words on epistemology were penned. Currently, most of those engaged in "science" accept a position best articulated by Sir Karl Popper.[52] Selection between competing theories is to be accomplished by appealing to a repeatable process called "falsification." This process does not determine whether something is true, but rather whether it is false; this, Popper claims, is after all positive knowledge of reality. Even though the inductive methods of science can never verify truth, they can falsify theory, and that might just be good enough. After all, if a theory has been "falsified," it can be discarded, while a theory that has passed "many" such attempts at falsifiability must be, by assumption, "true" or at least a "better" theory than the ones that had to be discarded.[53] Popper actually claimed to have solved the inductive problem itself.[54] Popperian falsification has been widely advocated, occasionally used, and brutally criticized,[55] yet it remains the process that most economists believe is consistent with "doing science."

All economics proceeds deductively. How, then, can one make decisions (that is, judgments) between the claims of competing theories? Bruce Caldwell put the matter succinctly when he wrote:

> Their choices, it seems, are twofold. They could attempt to criticize, and hopefully discredit, all rival systems. Such a task not only would be time consuming, it would be literally neverending. The second alternative, and one that appears more reasonable, is to propose certain criteria which could be used to critically evaluate their own and other such systems.[56]

At first glance, this "solution" of having decision-rule criteria, put forward by a physicist and follower of Popper, would seem to settle the issue of theory choice. Austrians might counter, however, that no validation of theory is possible by the application of empirical methods.[57] On the other side of the argument are the extreme positivists (or instrumentalists) who claim that the only purpose a theory has is to make accurate predictions. The more accurately a "theory" predicts, the better the theory is likely to be, both from a pragmatic viewpoint ("usefulness") and from the point of view of theory choice. It is the second contention that would trouble Austrians as well as others. The more predictively accurate theory must be, on these terms, the "better" theory. But why? Certainly not from the standpoint of *verstehen*—that is, understanding. And (Austrians might well add) if prediction is to be the test of a theory's correctness, then the entire edifice of modern macroeconomic modeling has to go, since the predictive accuracy of even the most sophisticated macroeconomic models (i.e., those with the most, and most difficult, equations) suggests that they continue to require reworking in light of past failures. Many people underestimate the errors generated by these models, because they focus on the actual prediction as against the official statistic and then take the difference between them as error. However, this ignores the *range* of potential errors. The economy seldom moves dramatically; it usually moves within a very narrow range. That constitutes the entire predictive range of macroeconomic models so that, when they miscalculate GNP growth by "one percent," they have missed by a very wide error indeed.[58]

It is important to understand why the Austrians do not accept the possibility of empirical "verification." An illustration from the physical world might help to illuminate their position. One of the linchpins of what came to be known as Keynesianism is the Phillips Curve. The idea behind it is that there is a "trade-off" between inflation and unemployment that is stable, predictable, and controllable. (In fairness to Keynes, he did not develop this analysis, although his work can be read as endorsing such a contention. In other words, had he lived long enough, it is questionable that he would have repudiated the incorporation of this tool into the Keynesian paradigm's box of such tools.) After Phillips did the original analysis, Keynesians all over the world began arguing that this empirical relationship in Britain would also be found in other economies. They set about doing "measurements" of such alleged relationships, claiming that full employment could always be "purchased" with just the "right" amount of inflation; if such a choice had to be made, they argued, surely every rational person would choose to suffer inflation to help out the unemployed? This "theory" was dutifully enunciated by politicians, taught to an entire generation of students (many of whom became policymakers themselves or went into the media, where they could explain this grand theory to average citizens), and incorporated into every standard macroeconomics textbook. Assurances were given that, contrary to the alleged theoretical possibility of rising unemployment and inflation, such movements were an empirical impossibility. The late 1970s brought an abrupt end to all this erroneous theorizing when the "trade-off" began to shift around and the economy experienced both high inflation and high unemployment (as critics of the Phillips Curve, among them the prescient and tenacious Austrians, had always warned that it could).[59]

The story of the misuse of Phillips's work is instructive on several levels. First, his original paper was simply a correlation of two data sets across a long time period. Even if those data observations had been accurate, nothing in the relationship implied either a causal factor or any applicability to any other data sets. Yet the economics profession immediately responded as though Phillips had put forth a *theory* instead of what was merely an empirical correlation. It was the inductive problem all over again—the impossibility of generating absolute truth from inductive processes. Yet the exaggerated claims of economists led them to forget this truth, and their distress when empirical reality destroyed the Phillips Curve was indeed great and well-earned. The Phillips analysis, far from being relegated to the history of economic thought, is still taught in macroeconomics as (with some minor qualifications) demonstrating a "generally true relationship" in macroeconomics.[60]

The story of the rise and fall of the Phillips Curve is not, of course, the only such development in modern economics that belies "falsifiability" as an operative (rather than stated) procedure. After all, Austrians had argued from as early as the 1930s[61] that short-run employment gains during inflations were always temporary and that the end result would be increased unemployment. Anyone who was familiar with the history, for example, of the Mississippi Bubble[62] could have argued in similar fashion. Inflation cannot create less unemployment but only, in the long run, more. To read today's economic textbooks, one would think that, to the extent that reality had "falsified" something, it certainly was not Phillips's correlation. The same can be said, of course, for the "failure" of reality to conform to some other theories in economics. One such theory is the Heckscher-Olin Theorem, a wonderfully plausible *a priori* construct that just doesn't seem to work out empirically. Nonetheless, it has not been discarded by the profession. The famous Cobb-Douglas production function is a somewhat different case. There is a wealth of "confirming" evidence, but some economists find it trivial and meaningless,[63] arguing that the usual specification guarantees such confirmations. These theories, falsified or not, continue to be used in empirical work; none is in danger of being discarded. Beyond that, economists avoid asking themselves some really tough questions—e.g., if they really believe their own methodology, then how was the Keynesian revolution possible? Where were the carefully thought-out and planned experiments that convinced so many of the profession of the truth of Keynes's unfalsified contentions? Naturally, there was no such empirical evidence beyond the casual observations of the depression years.

It cannot be the case that scientific procedures require the elimination of theories that have counter-evidence against them, or theories that do not make accurate predictions; if that were true, no theories would ever pass muster. Neither Malthus's theory of population nor Darwin's process of natural selection as a mechanism for "explaining" evolution can be empirically "falsified." Yet we do not discard them as non-science. Neither theory makes predictions concerning as-yet-unobserved states, so even instrumentalists should reject them—but, of course, they do not. In fact, since no theory can ever be proved true, because of the probabilistic interaction of the perceiver with that which is perceived, all such empirical tests generate information whose meaning is, at best, problematic and often simply a matter of opinion. What is actually tested in these statistical undertakings is an interesting question in itself, but theoretical statistics are not falsifiable—it may well be the statistical theory itself that is tested rather than the actual specified hypotheses.

Presumably, economists would like to say "meaningful" things about the world. They would like to believe that the things they say and personally believe are, in some sense, more rigorous than the things stated by laymen about the economy. Where, however, is this alleged superiority to be found, and how is it to be verified? Certainly not in the superior predictions that economists make by utilizing their theories, because their record is not enviable in that regard. Since economists are often wrong, agree about very little, and spend a great deal of time attacking each other's theories, what is the basis for their implied (and explicit) claims to superior economic understanding?

If we are all not to become nihilists, nor embrace Humean skepticism as the only correct methodology,[64] then how can we decide which theories are true and simultaneously demonstrate the superiority of the economic method? One answer, and it is the Austrian answer, would be to assert that there are two methods in science, one applicable to natural science and one to social science. This is called "methodological dualism."[65]

Competing social scientific theories are judged heuristically, both by the correctness of their assumptions and by the logical process of their deductions from those assumptions. Those theories most solidly grounded in real relationships will endure since, in the long run, they will not only produce accurate qualitative predictions, but will convince by the force of their logical purity and completeness.[66] For the person trained in today's positivist-influenced academic environment, this will not at first be a satisfactory alternative to empirical falsificationism. Yet, historically, this is precisely how economic theories have evolved, through the endless give-and-take that persuaded some individual minds that some theories are "better" than others.[67]

"Better" here does not mean "better" in the empiricist's sense. However, since economic empiricists don't throw out theories that have been falsified, they show through demonstrated preference (see Chapter Three) that, as a practical matter, theory choice based upon Austrian contentions is really not a radical idea at all but indeed characterizes most social theory choice and change. Lavoie has stated this position well:

> The problem of theory choice can be "solved" not algorithmically but intersubjectively; not logically within one mind, but dialogically among several. Adherents of alternative interpretive frameworks must endeavor to make their statements more intelligible to one another, to interpret one another's meaning, to pose one another's problems, *to persuade one another.*[66]

This, of course, will not be a satisfactory way to proceed as far as empiricists are concerned. The question is: what conflicts in economics have ever been satisfied by empiricist methods? Have we not, to echo Richard Weaver's lament, become obsessed with induction?[69] Or, as the late G. Warren Nutter wrote, in his discussion of that naive and peculiar faith that empiricists champion that looks forward to the day when all analysis will be positively grounded:

> there is even recurrent talk of establishing a purely objective, value-free theory of economic welfare, which would tell us how to make society better off while maintaining ethical neutrality.
>
> It is sufficient to say that all such thinking is wrong. Economics cannot be purged of moral content if it is to be concerned with the question of welfare...economics can escape moralizing but it cannot escape morals.[70]

Making society "better off" while maintaining "ethical neutrality" is precisely what both risk assessment and cost-benefit analysis claim to do. While it is sufficient to say that such thinking is simply wrong, it is not a simple matter to understand why such an argument can be so easily defended today. A careful examination of the arguments put forth throughout this book provides a basis for understanding why Nutter's claim may still be valid.

The objectivist-subjectivist disputes in economics and, for that matter, all science have profound effects upon the way in which we view and attempt to manipulate our world by the implementing of policy. Chapters One and Three examine two types of knowledge models and how they are used within a justificatory framework to defend policy choice. Both cost-benefit analysis and risk assessment methods display the spillover effects of the objective-subjective schism in economics. The purpose of the next two chapters is to demonstrate that subjectivity cannot be eliminated from either cost-benefit analysis or risk assessment technique and cannot, therefore, be eliminated from policy choice models. These techniques *can* be used—and generally *are* used—to add weight in favor of policy choices arrived at by other means. Yet many policy choices may not benefit at all from these techniques even when, *a priori*, it seems intuitively obvious that they theoretically can play a major role in public policy alternative decisions. Chapter Four, however, demonstrates that this is not always possible. Before proceeding with actual policy evaluation, there is one more theoretical hurdle to surmount. The next chapter provides an overview of cost-benefit analysis, and also provides an explicitly Austrian (subjectivist) critique of this common decision method/justificatory framework.

It needs to be remembered that, in many cases, theory choice *is* tantamount to policy choice and, even more strongly, policy choices imply theory choices as well. The process of defining policies requires a theory. Once these policies are defined, choosing between them is impossible without theory choice. Even rolling a die implies a theoretical choice construct— e.g., "randomness in policy choice is as good as or better than other methods of choice," or, alternatively, "randomness is worse than other methods for the choosing of policy and, as I favor bad policy decisions, let the die be tossed." There is simply no escaping the need for theory choice. This leads inexorably to debates on issues related to claims about human knowledge and actual reality. The old but unresolved debate over subjectivity as against objectivity in science forms the core for the analyses in the next two chapters.

3

A Subjectivist Evaluation
of Cost-Benefit Analysis

> *Cynics [cost-benefit analysts?] know the price of everything and the value of nothing.*
>
> Oscar Wilde
> *Lady Windamere's Fan*

> .... *sanity is not statistical.*
>
> George Orwell
> *Nineteen Eighty-Four*

This chapter will examine the theory and practice of cost-benefit analysis (CBA) in order to explore areas of disagreement—about such matters as technique, the interpretation of results, and general philosophical implications of the basic cost-benefit apparatus. This overview and examination is crucially important, since policy choices currently often require the elaborate trappings of cost-benefit justification.[1]

Cost-benefit analysis and Neoclassical economics are symbiotically bound together by the fact that cost-benefit analysis could not be done without the foundation provided by Neoclassical methods. When analysts measure, define, and aggregate "costs" and "benefits," they are doing so within the Neoclassical framework. It is fair to argue, therefore, that cost-benefit analysis is the "child" and Neoclassical economic theory the "father." Neoclassical economics can stand alone without CBA, but CBA falls without Neoclassical economics. It is necessary, then, to understand basic Neoclassical contentions before one can understand CBA.

Neoclassical Microeconomic Theory: Building Blocks of CBA

Neoclassical economic theory stresses several interrelated foundational pillars. Among the most important are:

39

(a) logical rigor,

(b) resource scarcity,

(c) perfect information (in many cases),

(d) quantifiable methodology (for empirical testing), and

(e) efficiency (both economic and technical).

No one person speaks for the entire paradigm, so the listing above is a generalized construction. These characteristics should be combined with the analysis which follows. Those familiar with the terrain will probably not have substantial criticisms of the list. It can be inferred from the above listing that economists hold conceptions about both the theoretical world (logical rigor, perfect information, quantifiable analytics) and the physical world (scarce resources, efficiency outcomes). An exhaustive description of these conceptions is not necessary for this discussion, but a further set of generally agreed upon methodological propositions will be useful. Hollis and Nell provide just such a compilation:

(a) claims to knowledge of the physical world can be justified only by experience

(b) whatever is known by experience could have been otherwise

(c) all cognitively meaningful statements are either analytic (true by definition) or synthetic (empirical fact) but not both

(d) synthetic statements, being refutable, cannot be known "a priori"

(e) analytic statements have no factual content

(f) causal laws are simply well-confirmed empirical hypotheses

(g) the "test" of theory is the accuracy of its predictions

(h) analytic truths are true by definition

(i) judgments of value have no place in science

(j) science is distinguished by its subject matter, not by its methodology[2]

This is a fair summation of the empiricist approach generally taken by modern Neoclassical theory. Although individual economists might wish to eliminate or revise certain parts of the list, the summation is adequate for our purposes in this chapter.[3] To the above list, Hollis and Nell added the proposition known to economists as "methodological individualism."[4] This is the doctrine that the values and actions of individuals are the proper focus of study in economic theory. As applied to CBA procedures, this means that it is the values, costs, and benefits that accrue to (or fall on) individuals that most often must be taken into account. This leads to an important consideration in CBA known as "social discounting." This issue will be carefully explored in a later section.

The Neoclassical paradigm combines a stated empirical approach to economics with other assumptions, two of which are of primary importance for performing cost-benefit calculations. The first assumption is that pervasive scarcity exists in the physical world. The second is that various "states of the world" (known in economics as welfare states) can be estimated, compared, and evaluated. Neoclassical analysis proceeds on the assumption that the task of all economic arrangements, whether market or social, is to allocate scarce resources so as to maximize individual welfare.[5]

This position concerning scarcity and public policy is sometimes generalized into the belief that legal arrangements as well as policy should be designed to maximize societal wealth.[6] Since all resources that are not free gifts of nature have a positive price, the fact of scarcity can be deduced from the existence of price. "Cost," therefore, is something that all planners, public or private, must and do face and should take into account.

Further, the concept "scarcity" also logically implies considerations of "efficiency." In the absence of scarcity, all efficiency questions would be meaningless—for then output would not be constrained, and people could have as much of any good as they wished. What, then, is "efficiency"? Technical efficiency, in Neoclassical terms, is defined as obtaining some given output with minimum inputs. Economic efficiency is defined as producing a given output while simultaneously minimizing total input cost. Can there ever be conflict between efficiency criteria and CBA judgments? Not if a particular policy passes a CBA test, as well as a stringent welfare test; it is then said to be efficient and desirable. The most stringent welfare test was formulated by Vilfredo Pareto, and is discussed at length later in this chapter.

Neoclassical CBA occurs within a game-theoretic framework based upon unchanging rules and assumptions.[7] The logical framework is called perfect competition theory. The more closely resource transfers approximate the assumptions of perfect competition, the greater the chance that these transfers enhance welfare as defined by the paradigm. Further, a concomitant of perfect competition doctrine is that the theory approximates the true opportunity cost(s) of resources.[8] For this reason, such outcomes are, by definition, "efficient" within the paradigm's definitional framework. Such approximations are always tenuous because the assumptions of perfect competition obscure actual competitive processes in the physical world.[9]

A perfectly competitive market is characterized by:

(1) perfect information,
(2) perfectly uniform products,

(3) many small firms, none having market power,[10]

(4) no advertising or brand names,

(5) perfect mobility of resources, and

(6) freedom of firm market entry and/or exit.

Under such assumptions, the price of the final output will, in the Neoclassical view, closely approximate the true opportunity costs of production. Further, assuming that demand for this output is determined by consumers on the basis of marginal valuations taken as axiomatic,[11] all efficiency criteria are then satisfied. The system is completely *a priori*, self-contained, internally consistent, and logically rigorous. Further, it is manipulable by the techniques of calculus and geometrics, a pair of methods the Neoclassical economists prefer to use since, they argue, these methods are much more rigorous and scientific than rhetorical theorizing.[12] The actual competitive process that the theory seeks to "predict" is really very different from the sterile world of competitive theory. In the real world, competition is imbued with uncertainty which leads to dynamic change, imperfect knowledge, and other model violations. The perfectly competitive model assumes these away or, to use the terminology common in economics, it "abstracts" from them. At base, however, the theory is used both as a predictor of real-world events, and as a standard by which actual outcomes are to be judged. The statement "this is not a perfectly competitive market" makes sense *only if* this is assumed to be of consequence in some theoretical (normative) way by the Neoclassical theorist.[13]

Since the perfect competition model is easily adapted to mathematical treatment (and, hence, symbolic manipulation), it is—at least in theory—possible to use it for empirical measurements. All CBA studies require measurements of a specific kind whose validity is argued for by the analyst, as long as a particular range of error is admitted. It is possible to use both geometrical illustration in order to "see" resource misallocation and mathematical specification in order to measure such misallocation.

Figure 3-1 represents a firm that has some monopoly power. This firm is not in a perfectly competitive market, nor does it behave as a perfectly competitive firm should theoretically behave. According to the behavioral assumptions of the perfect competition model, business firms act so as to maximize profits, which this firm does by setting output at the point where its marginal cost of production equals the marginal revenue it receives for that output. This can be seen as output level Q_m in Figure 3-1. The price the firm will receive for output level Q_m can be read directly from the demand curve; it is P_m. Because this firm has monopoly power, the selling price does not equal the marginal cost of production; rather, it is higher by the distance OP_m minus P_{mc}.

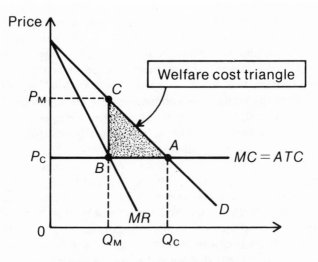

(Figure 3-1: Standard Monopoly)

The loss of "consumer surplus"[14] can be seen as the triangle *ABC*.
Neoclassical theory implies that this is a "bad" outcome, because there
is "inefficiency" due to resource misuse by the imperfectly competitive
firm. It is not necessary to explore the reasoning at this point, but (in theory,
anyway) the triangle *ABC* could be empirically measurable.

A perfectly competitive firm in long-run equilibrium would be dia-
grammed as in Figure 3-2 as follows:

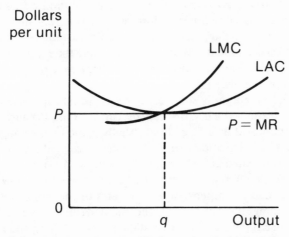

(Figure 3-2: Long Run Equilibrium of a Perfectly Competitive Firm)

This competitive firm produces output level Q_c because that is the level that equates marginal revenue and marginal cost. Once again, the price is given by the demand curve at Q_c; it is P_c. The price at which the output is sold to consumers just equals the cost of producing that output at the margin. Since a demand curve can also be viewed as a collective marginal valuation function,[15] there is no "welfare loss" in this market as there was in the imperfectly competitive case above (Figure 3-1). The key difference is that, although both types of firms equate marginal costs with marginal revenues, competitive firms actually sell that output at a price that equals their marginal costs, while firms with monopoly power can typically sell at a price that is above the marginal cost of production.

All perfectly competitive firms are "price-takers"; that is, they respond mechanically to given data without having any control over it. The reason this is an issue in CBA is that the benefits and/or costs associated with any undertaking are going to depend to some degree on the competitiveness or noncompetitiveness of the institutions supplying the economic goods in question. To the extent that the actual prices diverge from those that would be set in perfectly competitive markets, adjustments have to be made in the CBA process.[16] Although this sounds rather simple, in practice it is virtually all guesswork. In many cases, there are no prices whatsoever to use as actual data, and shadow prices have to be estimated in order to complete the analysis.[17]

The Neoclassical view of economic reality might be distilled to simple cases of constrained maximization within given data sets. Model consumers maximize "utility," subject to a budgetary constraint. Firms maximize "profits," subject to the twin constraints on production of input price and the "state" of technology. Even in those cases where solution values cannot be known, signs can sometimes be inferred.[18] It is hardly surprising that techniques like these that have been so useful in nineteenth- and twentieth-century science should have been imported into the social sciences as well. Yet the Neoclassical view of the world of economic reality is but one of many competing visions.[19]

The Neoclassical paradigm is singular, however, in that it is the only economic system within which CBA can be carried out. Because this paradigm treats the world as a large (and usually static or unchanging) information set, the calculations of CBA can be carried out cross-sectionally and extrapolated across future time frames. Yet since the majority of data on which CBA studies are based change over time, how can the results ever be made commensurate with states of the world not yet extant, even assuming the validity of the method? Even within the Neoclassical view itself, the cross-sectional data is only a fleeting and transitory artifact that is in the process of changing as firms and individuals act so as to fulfill the behavioral assumptions of the model. It therefore cannot be true that cross-sectional data can be correctly used as proxies for the "values" of consumers, or as actual "costs" for firms carrying out productive processes in even the present period, let alone in all future periods. Historical data sets are merely the equivalent of finding the "broken artifacts" associated with some "lost civilization."[20]

In summary, the Neoclassical methods and models tend to generate outcomes already contained within the structure of the assumptions of the theory. When the world is viewed through the Neoclassical lens, it resembles an unchanging "data set" overlaying a series of constant relationships. When the data are inserted into the models, the Neoclassical approach generates answers that confirm its assumptions about economic reality. Yet reality is more varied and much less systematically comprehensible than Neoclassical economic theory tends to suggest.

Cost-Benefit Technique: An Overview

Cost-benefit analysis can be viewed as the applied use of the Neoclassical economic paradigm in order to answer the questions raised when public policy is implemented. At the very least, CBA requires the (fairly) accurate measurement of three key variables: costs, benefits, and the so-called social rate of discount—i.e., the interest rate at which alleged future benefits are to be discounted to the present. The basic evaluative formula is:

$$P.V. = \sum_{t=0}^{t=n} \frac{B_t}{(1+r)^t}$$

Where

B_t = benefit in period t

r = discount rate

n = the useful life of the project under consideration

If all this information is known, the above formula will show net benefits (net costs) for any considered project. The analysis itself ideally should be *wertfrei* or value-free. The question of whether or not a particular policy "ought" to be undertaken has nothing to do with the procedure itself, except insofar as the social rate of discount is varied to account for the fact that it is a public endeavor. That singular caveat is of the utmost importance for CBA, since it is usually argued by CBA analysts that this discount rate should be set lower than prevailing private market-generated discount rates. Lower discounting will increase the likelihood that net benefits will be shown by the technique. For that reason alone, the computation of the social discount rate merits the most rigorous scrutiny.[21]

Not every project that produces a CBA net benefits result *ought* to be carried out, let alone publicly funded and administered. For example, air traffic control has been shown to have net benefits, yet it need not be administered by the government.[22] If the requisite information is known, it can be demonstrated that net benefits will accrue to consumers of air travel if traffic control is in place. But the analysis would be purely "positive." It would be a fact, but facts cannot tell us whether air traffic control *ought* to be provided by government or can be left within the private sector. All such controversies ultimately rely on a foundation that must be supplied by political philosophy. The *ought* issue, really, is about legitimate state actions.[23] Economists *qua* economists (or as CBA analysts) cannot decide these sorts of issues from within the positive analyses of their science. The final decision to implement or not to implement the analyzed policy is independent of the CBA's final determination.[24]

Judgments such as the one discussed above are not exclusively about questions of economic efficiency. A project that CBA has shown to result in negative net benefits might nonetheless be retained or implemented for purely political reasons. There is also the built-in pressure that lobbying creates around ongoing public policy. In addition, there is the possibility that the program, even though it demonstrates net positive benefits, "ought" to be carried out by the private sector—either for reasons of efficiency or because the program would breach some morally agreed-upon limit to the proper activities of the government. "Privatization" has been an attempt to transfer valuable public programs into the private sector, sometimes returning them and sometimes injecting existing public programs of value into the private sector for the first time. The question of whether or not these changes "ought" to be made is not one that can be solved by the application of CBA.[25]

CBA Assumptions Concerning Actual Variable Measurement

Before CBA can be operationalized, the key concepts "cost," "benefit," and "social discount rate" must be carefully defined. Obviously, how these concepts are defined is vitally important in determining the magnitudes that any CBA will generate. Any incorrect specification of these variables will inexorably lead to illegitimate results just as surely as would, for example, simple data error.

"Costs," for the Neoclassical economist, are always "opportunity costs" in theory, but actual existing (or estimated) money costs in CBA practice.

In economic theory, the costs of resource X are the forgone opportunities $(Y_1, Y_2, \ldots \ldots Y_n)$. In practice, since the alternative set (and, therefore, its "costs") cannot be known, the CBA has to use money prices, or so-called "objective" costs. The Neoclassical economists argue that the closer markets are to being perfectly competitive, the closer actual observed money costs will be to real opportunity costs. In fact, in perfectly competitive markets, costs and benefits are neatly defined and measured, as (by construction) the money costs in such a market equal the opportunity costs of production and the benefits that consumers derive from that production.[26] When the set of possible alternatives can be narrowly defined, then the cost of any action is the next-highest-valued opportunity sacrificed in order to take a particular action. The problem here, from a subjectivist point of view, is precisely that this can never be known *ex post;* therefore, neither can true opportunity cost.[27]

Cost-benefit analysts are not in agreement on the question of the correct rate of social discount. First, it should be noted that "social" entities are not the usual unit of analysis in Neoclassical economics. The operating framework is termed "methodological individualism" because individuals, not groups, are the subject of analysis. The social discount rate is some single number that is supposed to reflect the collective time preference rates accurately for public undertakings on the part of millions of individuals whose actual preferences can only be estimated. Second, as mentioned previously, the choice of that rate has a large effect on whether or not the project under consideration will show net benefits. This concept, then, has provided a useful and controversial arena for discussion in both CBA and economic theory itself.[28]

One problem with social discount rate calculations is that basing them on internal rate-of-return data can be either misleading or impossible.[29] Consider the following:

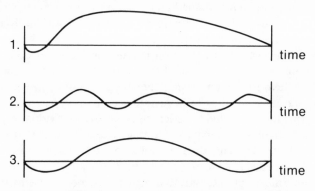

(Figure 3-3: Rates of Return Under Varying Assumptions)

In Figure 3-3, the second and third public projects will not show a consistent rate of return. In the second project, there are periodic expenditures for the replacement of major equipment. In the third, there are major disengagement costs (for example, as with the decommissioning of a nuclear power plant). A single comparison of the internal rate of return on these three projects would be misleading. This had led some economists to suggest that, when these sorts of cases arise, a different measure of benefits should be employed.[30]

On those occasions when an internal rate of return can be estimated, at what level ought the discount rate be set? One contention is that it ideally should reflect the opportunity cost of using capital in the private sector. The reason given is that it is precisely private capital that is taxed away for public uses, so that the true opportunity cost must be the alternative to which it would have been put. This assumes, of course, that the project is not funded by the monetization of debt, but from current tax revenues.

Although this seems both logical and obvious, it simply is not feasible from a practical standpoint. Several complications arise that make the calculation less than completely accurate. First, there is the effect of the corporate income tax, which means that net (rather than gross) yields need to be calculated. Second, there is the fact that virtually all markets are, to varying degrees, imperfectly competitive when analyzed from the Neoclassical viewpoint. Therefore, some reduction of rates of return must be made in order to allow for the "monopoly power" exercised by some firms in the economy. Finally, some economists have argued that, since the corporate tax is a "given," a social project that pays less than a private one might still be preferable if the private project was one that produced no tax monies. But others have suggested that higher rates of return that are often argued for in public undertakings are *themselves* merely a result of the distortive effect of the tax, lowering net private rates of return.[31]

Arguably, it can be maintained that attempts to use private market discount rates incur serious calculational difficulties. Alternatively, some economists contend that private market rates should not be used as the basis for public expenditure decision-making because they do not reflect "social preferences." They argue that the private interests of individuals often blind them to the long-term benefits that a public project will produce, generating what they term private sector "myopia."[32]

Under this theory, individuals left to their own individual preferences would fail to plan adequately for future generations' needs. If this argument is accepted as valid, then some income transfer forward ought to occur and, for that reason, the social discount rate should be set lower than the alternative private rate. Yet, as logically persuasive as this argument might be, it raises difficult practical issues. The analyst immediately encounters what could be termed the "stopping-point problem." Where exactly (and logically/empirically within the economic method) does the analyst stop lowering the rate? At lower rates, more projects will show net benefits, as they will if time horizons are extended. How far ought income to be transferred forward, and at what discount rate? This theory presents the analyst with no consistently defensible answer. In fact, the number of positive-benefit public undertakings that can potentially be justified under this view might well outrun the ability of the government to finance them and will likely intrude into areas where the private sector is already performing well.

Another related argument is that the social discount rate ought to be set lower than the private rate because public sector undertakings are less "risky," and since risk is a component in most interest rates actually charged in the market, this lower risk justifies the lower rate.[33] The argument is that the defaults, though they can be costly, can be "socialized" across the entire universe of taxpayers (currently and in the future), so that government projects will be viewed as less risky by investors. The recent savings and loan tax revenue increase proposed by President Bush is an example of the public sector's ability to raise vast amounts of money from taxpayers when necessary.

Yet, though this argument makes sense when both projects and adverse conditions are few in number and magnitude, it begins to lose some of its force when the government begins to become overextended in terms of its own balance sheet—or when the emergencies become frequent or large. It can only be assessed from within a particular existing project universe and, judging from budgetary constraints and the increasing number of bailouts required from the government, its current applicability is narrow—if it has not already shrunk to nothing.[34]

One suggestion for solving the stopping-point issue and setting the social discount rate was Harberger's:

> The fact that the social rate of discount may lie below the marginal productivity of capital proves only that the rate of investment should be expanded; *it does not prove* that, for a given rate of investment, capital should have different marginal rates of productivity in the public and private sectors. The end result of an optimal investment policy would therefore be a situation in which the marginal productivity of capital in both the private and public sectors was equal to the social rate of discount.[35]

Given that, by assumption, the social rate may lie below the private rate, there are still problems with the actual calculation of these rates of return. One such problem is that the social rate may change over the life of the project under consideration. In this case (and assuming that we knew what it was at first and how it is changing), the original formula can be modified as follows:

$$P.V. = \sum_{\substack{i=1 \\ \Pi}}^{\substack{t \\ }} \frac{N_t}{(1+ri)}$$

As Harberger writes:

> It is unfortunate that the great bulk of the literature on cost-benefit analysis has been based on the simplifying assumption of a constant discount rate...[36]

Choosing the Social Rate of Discount

Cost-benefit analysis is done in a dynamic universe where there are no constants.[37] Social discounting is a necessary part of CBA not only because people typically discount the future, but also because it is the only constituent part of the analysis that attempts the operationalization into dynamic terms of what basically is a purely static procedure. The problems for CBA calculations caused by the passing of time are myriad and subtle. All static models freeze time, moving from one timeless state to future timeless states where the actual process of movement is simply assumed away or given plausible rhetorical story-telling as support.[38]

Individual valuations change through time, which is why people have differing *ex ante* and *ex post* valuations of past decisions. This is a major problem for any theory which seeks a single, time-frozen preference answer for policy costs and benefits. But without the assumption of unchanging preference scales through time, CBA results, at best, would be outdated before a project could ever be finished. (At worst, they would simply be meaningless and therefore totally irrelevant.) Yet we know that people's preferences do not remain constant through time. For this reason, even if we assume that all data is accurate now and that the CBA provides a perfect method for assessing policy, there would be no reason to argue that the *ex ante* conclusions of that study would still be valid at some point in the future.

Time is unambiguously unidirectional and dynamic, and yet all cost-benefit studies are cross-sectional, the data frozen at some instant.[39] Conclusions about future social states are to be made on the basis of historical data sets, or data that is simply invented for the purpose at hand. Adjustments to current and projected data must be made because of the universal human tendency to discount the future. The choice of this rate of discount, as has been discussed just above, is of crucial importance for the analysts. If that rate is set low enough, virtually any conceivable project might show net benefits after the standard formula is applied. Given this fact, it is not surprising that much debate has focused precisely on the question of the appropriate rate of social discount, and much criticism has been directed at public agencies such as the Army Corps of Engineers for their use over the years of abnormally low rates of social discount.[40]

Part of the magnitude of interest rates is an allowance for risk. All investments involve some sort of risk, and the more risk there is, the higher the interest rate is likely to be set for that investment. In a world without risk, no risk assessment would be necessary; the task of the cost-benefit analyst would be appreciably easier, but still impossible to accomplish with certain accuracy. Austrians do not agree that using either the current private discount rate or using some different and lower social rate is an acceptable alternative in CBA. Eliminating risk does not eliminate the discounting problem, because, for Austrians, interest rates in pure theory are a function of time preference alone. Since interest is not risk-caused, the discount problem would remain even if all risk were eliminated.[41]

The objection against using the private market rate, even if there were a uniform discount premium, can be stated as follows: the opportunity cost of capital (its "marginal efficiency") cannot be gleaned from a simple observation of private markets, any more than the correct opportunity cost of any action can be so observed. Further, there is no one private rate of discount. In fact, there is a whole panoply of rates, each one representing different appraisals of the nature, size, and time-frame adhering to certain transactions.

Precisely which rate, then, ought to be used as the proxy for the private rate of discount? The CBA often takes the average of some collection of such rates, or simply uses the current Federal Funds rate as the appropriate proxy. Alternatively, an average rate can be computed by using those rates prevailing in the capital markets from which the confiscated capital is to be drawn. This is done in an attempt to approximate the opportunity cost of that capital. Each of these procedures has its own set of conceptual/procedural difficulties.

An average of major rates would tend to be relevant to the extent that the riskiness of the project mirrored that of the projects from which the rates were taken. This would happen only by pure chance, even assuming that the actual "riskiness" of public projects can be calculated accurately at all *ex ante*. The Federal Funds rate[42] (and, for similar reasons, the so-called prime rate) will tend to understate the risk component since these rates *assume* less riskiness than the rates associated with most private undertakings.

A combined rate that uses a weighted average of the rates prevailing in the markets from which the funds are to be drawn looks good in theory. As a practical matter, however, it is generally impossible to ascertain the origin of taxed capital. Further, unless this capital is voluntarily made available to the government, these rates will not accurately reflect the true opportunity cost of the capital. Capital that is coercively taken from individuals should be discounted at a rate that is higher than the prevailing market rate in order to allow for the fact that it is being used for less valued alternatives. The public project is (at least for them individually) a worse alternative than their privately valued use option(s). Some premium should attach to these private rates, then, in order to take into consideration the origin of the capital. The Neoclassical response will be discussed below.

What about the argument that public undertakings are inherently less risky than private ones, and therefore that the social rate of discount ought to be set below the private market rate(s)? While this might be true for the first few projects undertaken by the public sector, the force of this argument becomes much smaller as the number of projects becomes ever larger.[43] Historically, many governments have defaulted on their obligations; many more will in the future. The thought that our government might join future public defaulters has now become seriously entertained.[44] The historical record, as well as the current concern over the fiscal health of the American government itself, suggests that the power of governments to socialize costs and risks might well be limited, and that we are reaching and sometimes risk surpassing those limits.

Preexisting distortions in discount rates caused by prior macroeconomic intervention is another troubling problem for CBA. There is no reason to believe that the accuracy of private rate(s) is somehow "correct." Expectations about what the government might do "soon," as well as evaluations of the effects of what it has already done, are going to affect discount rates that prevail cross-sectionally. This is going to "confound" attempts to use these rates as accurate reflectors of private discounts, since these perceptions change daily. Interest rates now fluctuate in a single day more than they used to change over years.[45] It might simply be impossible to speak of a "private rate" at all, except in the sense of Wicksell's "natural rate" or von Mises's "originary rate," both of which are purely theoretical constructs.[46]

The Neoclassical rejoinder to the problem of capital coercion is to invoke "public goods theory." Taxation, for Neoclassicals, is not really a manifestation of lower valuations by people for governmental projects, but is rather the outcome of the "free-rider problem" that attaches to the provision of all public goods.[47] Under this assumption, it is quite possible to generate net positive benefits while using confiscated capital, because

individuals—though they actually desire such expenditures themselves—
have no real incentive to make this desire known to the taxing authority.
They have no such incentive because they each believe that someone else
will pay for the project anyway, and that they will be able to have a "free
ride" on the outcome. For the Neoclassical economist, then, the desirability
of public projects is simply another "given" that is more than adequately
explained by an appeal to public goods theory.

Austrians, however, would not accept this argument. For them, nothing
is "given" except valuations through individual actions, and an acceptance
of the free-rider doctrine would imply an abandonment of demonstrated
preferences.[48] If Neoclassical theory had an absolutely objective mechanism
for determining whether a good was indeed "public," and to what extent
it should therefore be governmentally provided and/or regulated, then this
argument would be as useful to all as it is compelling to some. Unfortunately, virtually anything can be argued to be a public good or to have public-
good aspects that require capital confiscations like the kind we have been
discussing. In light of this disagreement (and the lack of such a
demonstrable objective apparatus), Austrians remain unconvinced by this
line of argument. Virtually all human activities have "spillover" effects,
be they in terms of "costs" or "benefits." It can sometimes seem from
cursory readings in Neoclassical textbooks that each one of these effects
can be used as justification for some governmental intervention.[49]

Problems of Time

The bulk of the cost-benefit literature proceeds on the assumption that
predictions about policy are the same as actual policy outcomes. To be
accurate, cost-benefit analysts really need to know two sets of data, one
ex ante and the other *ex post*. In practice, they cannot know either perfectly,
and the *ex post* data can only be approximated by educated guesses. The
most sophisticated predictive models currently used in economics are
notoriously inaccurate, so those projections are likely to miss the actual
numbers by large amounts.[50] It might make a considerable difference were
follow-up studies done after policies had been implemented, if such studies
might eventually trigger policy termination. Once policy implementation
begins, the "sunk cost argument" is a powerful inducement to continue
even demonstratively erroneous undertakings. Public laws are seldom
repealed, public projects seldom abandoned. This accounts for the large
number of "emergency" laws still in operation, such as New York City's
rent control ordinances, the Department of Energy's "synfuels" program,
and the McFadden Act of 1927 prohibiting banks from branching across
state lines.

The cost-benefit analyst, then, faces a monumental task. Incomplete (and sometimes inaccurate) data must be adjusted because of the fact that markets are not Neoclassically perfect. Assumptions must be made concerning the alternative return to resources, both employed and unemployed.[51] A "correct" rate of social discount must be estimated; even if data is nonexistent or sketchy, future states must be predicted. On this basis, a decision is ultimately made by the analyst (but not necessarily followed by the public authority) concerning a new project that might produce effects different from the ones projected in the analysis.

The list of potential problems is not yet exhausted. A further possible complication is that the distribution of costs and benefits is an uneven process, not a static outcome; therefore, the issue of "standing" needs to be examined. Who (or what) has "standing" in CBA is the question of which values and distributional effects are to be viewed as important enough to be considered by the analyst when aggregating numbers. Social welfare theory is only an adjunct to standard theory, beginning with the utilitarian requirement that the greatest good for the greatest number was the rationale for policy. The problem with this rule is that it grants virtually unlimited power to majorities, no matter how small the majoritarian margin.

Social philosopher Vilfredo Pareto developed a very stringent welfare rule that has come to be known as Pareto-optimality. Under this rule, all changes that generate net improvement in at least one person's "welfare" without harming any other person are termed Pareto-superior moves, and are permissible. Once a social state is reached where it is no longer possible to make anyone better off without a concomitant worsening of at least one other person's welfare position, then Pareto-optimality has been reached and no further moves are permissible. This formulation exhibits at least two virtues: it is internally consistent with the Neoclassical paradigm[52] and it forbids the interpersonal comparison of subjective utility states (hence it is a limited and strict criterion by which to judge changes in conditions).

The Pareto rule had to be abandoned for two reasons, the first theoretical and the second pragmatic. First, it is operationally impossible to adhere to a Paretian restrictive social welfare rule: while the rule is valid within the context of pure theorizing, it is not possible to compare this analysis to reality. Second, as vast welfare states emerged in the Western democracies, the necessity of discarding so restrictive a guide for welfare policy became apparent within a short time. Pareto's formulation would not allow the overwhelming majority of public projects to be defended on Neoclassical welfare grounds. The easiest solution was, therefore, to find another rule. This search for another welfare criterion became essential after the Neoclassicals' declining marginal utility of money argument was refuted, since it could no longer be maintained that progressive taxation and income transfers were clearly good.[53]

The impossibility of comparing internal (and hence subjective) welfare states led Sir John Hicks and Nicholas Kaldor to formulate the Kaldor-Hicks criterion for policy evaluations. Under their rule, which is now widely accepted by both cost-benefit analysts and economists, any change between social state one and social state two is a net improvement (and therefore ought to occur) if the winners can compensate the losers and yet remain winners.[54] This relationship between the so-called compensation principle and CBA is direct, for the winners could only compensate the losers and remain winners if the benefits of the policy were greater than the costs.

Faced with the same impossibility of accurately measuring internal states, the Kaldor-Hicks criterion *does not require that such compensations actually occur.* If we were to attempt to find out who the winners and losers are by polling those citizens whom analysts believe are likely to be personally affected by the program, the winners would have too big an incentive to lie and proclaim themselves losers entitled to compensation. The rule brings the analyst no closer to a quantitative solution, but it does provide an intellectual defense of policy on grounds that Neoclassical economists can understand and appreciate.

There is a watered-down version of this rule that is often used in CBA. Instead of trying to measure internal microeconomic distributional effects that some policy might have, simply aggregate the effects and then make a macroeconomic argument in favor of the implementation of the policy. Total the benefits and costs of the project through the application of CBA. If the benefits exceed the costs, then the project is justified without winners compensating losers. There will be adverse distributional effects, but the macroeconomic effect is positive and, when examined from a utilitarian standpoint, the project will pass examination. The actual distributional effects are unknowable. Since no analyst has the means to measure the valuations that attach to resource use by individuals, this becomes the only way in which policy can be evaluated by Neoclassical theory.[55]

There is another way in which a solution might be found for this conundrum. If the assumption is made that the macroeconomic whole is, in some sense, distinct from and morally superior to the microeconomic parts, then the question of "what society wants" takes on actual meaning. For if society can be argued to have desires and a separate rate of social discount, then the analyst, assuming that there is a method for the determination of these objects, need not be concerned with the preferences and distributional effects upon individuals at all. This approach has been advocated, most notably by Bergson,[56] but it is replete with unsolvable problems.

Who speaks for "society"? How does this person (or these people) know what society wants, and at what discounted rate? One answer that can be advanced in democratic societies is that voters demonstrate through the voting process precisely the answers to these questions. Some public finance specialists have accepted this position, but cost-benefit analysts, by and large, have steered clear of formulating their decisions upon so weak a foundation.[57]

That foundation is weak for the following reasons:

(1) The majority of eligible citizens do not vote in the United States, even in presidential elections.

(2) A vote cast cannot be taken as support for the thing that is voted for— it might be a vote against something else.

(3) Voters are sometimes ignorant of what they are voting for.

(4) The lack of unanimity suggests no such thing as a "societal" choice.

(5) Arrow showed that outcomes through voting can violate the Neoclassical assumption of preference transitivity and, therefore, are not "rational."

(6) The outcomes would have to remain unchanged even if those not voting decided to vote.

(7) People seldom vote directly on policies at all.

One would then have to argue—assuming that all the above problems could be settled—that "society" actually got the programs they thought they were voting for through the political process. In other words, it is *not sufficient* to demonstrate that a process could conceivably validate the "society" approach to CBA; it must also be demonstrated that the process actually produces the results sought by those who participate in that process. It is clear, given these immense difficulties, why so many cost-benefit texts and so many analysts have put aside the collectivist approach to decision-making entailed by Bergson's formulations. Unfortunately, the Neoclassical paradigm has reached the limit of its ability to deal with welfare issues at all, for most of this analysis is quite dated and nothing new has come along to replace it.

Subjectivist Economics: An Austrian View of Cost-Benefit Analysis

Several alternative approaches in economic theory are distinct from the Neoclassical paradigm, among them the Marxist, Institutionalist, neo-Ricardian, and Austrian "schools."[58] Although Carl Menger, the father of the Austrian school, is usually credited with being one of the three co-discoverers of marginalist economics during the early 1870s, the road followed by the Austrians was quite different than that followed by the British Classical and Lausanne schools. The Austrians maintained a thoroughly subjectivist methodology, while the Neoclassical economists used marginalism as a means to mathematize and (they hoped) "objectify" economic science. This distinction between the objective and subjective approaches to reality has broad implications for all of science.[59]

The first major difference that one encounters when studying the disagreements between the two paradigms is over the nature of science as it applies to human action, and as to whether the economist should follow methodological monism or dualism. The Neoclassical economists of the late nineteenth century began importing into an extant body of essentially *a priori* theory an entirely different kind of theorizing borrowed from physics and Newtonian mechanics.[60] They did this because they saw parallelism across disciplinary boundaries. They believed in the power of that theory to produce useful results, and they desired the prestige that practitioners of those techniques enjoyed. But they were, according to Mirowski (1989), borrowing techniques without sufficient rhetorical (and metaphorical) foundations. Early attempts by physicists and mathematicians to elicit from economists a description of exactly what they thought they were doing were met with "incomprehension" or "silence."[61]

At the same time that economics was being transformed into a small branch of physics and mathematics, the practitioners themselves began to claim something quite remarkable about what they were really doing with their theory. By the third decade of the twentieth century, economists were beginning to argue—along with philosophical positivists—that only reality mattered, and that all these newly acquired mathematical techniques were necessary tools for economists to meaningfully talk about reality. There was a mass movement away from the great verbally deductive models of the past, such as those of Marx and Ricardo, even as the newer theorists erected great *mathematically* deductive models of their own.

To noneconomists, this development appeared (and appears today) bizarre. As increasingly complex theory piled on the already existing complex theory, economists began to proclaim that they were actually engaged in an empirical exercise, not a theoretical one. Further, they began to believe that these methods—borrowed, as they were, from nineteenth-century physics—were the *only* appropriate tools with which economists could study the economic world. The *locus classicus* of this position can be read in Milton Friedman's 1953 article on method in economics.[62]

One interesting aspect of the Austrian-Neoclassical disputes has been a charge leveled by each school against the other: "extreme *a priorism*."[63] Austrians accuse the Neoclassical school of erecting vast and purely theoretical artifices within which outcomes are predetermined by assumption. In this regard, Austrians attack perfect competition theory as static and meaningless. Conversely, Neoclassical theorists attack Austrians for creating their own self-contained models, as Marx did, that arrive at conclusions "unsupported" by empirical investigations. This dispute often pits the two schools against each other on methodological grounds, even when (as often happens) agreement has been secured on actual policy evaluation. These disputes cannot be completely catalogued in this book, but it should be noted that they have been both fruitful and interesting. Unfortunately, it is also true that they have sometimes been pursued with less that academic civility.[64] This part of Chapter Three will therefore focus on a narrow set of issues: where the Austrians would take issue with the Neoclassical definitions for variables that are both a requisite part of cost-benefit analysis and a continuing operational problem for it.

Disputes Over Definitions and Measurement: Costs

Cost-benefit technique requires objective data, not philosophical opportunity costs. Prices, demand functions, resource input costs, costs borne by taxpayers and third parties, benefits, discount rates—all these need to be given flesh before a CBA can be operationalized. But these numbers are, to use Hayek's phrase, the "facts of the social sciences." What if none of these things can be accurately known? Then, *ipso facto*, no CBA studies can take place by using existing methods. If this is true, then absolutely no claims can be advanced for the "rationality" or efficiency[65] of a particular public undertaking. In brief, the acceptance of Austrian contentions regarding the nature of these variables (concepts) entails the abandonment of the alleged scientific foundation for public policy altogether. Because, then, of the Austrian conceptual destruction of CBA, the Austrian critique of Neoclassical definitional outcomes is unlikely to be widely embraced by the profession in the near future. It is instructive, however, because it could be used to strengthen our economic analyses if only it were better understood by Neoclassical practitioners.

Austrians do not question the basic concept of opportunity cost; indeed, they extend the analysis somewhat further. Since the opportunity (or set of opportunities) is never pursued, there is simply no way that they can be compared *ex post*. That being true, there really is no actual cost, except in the mind of the individual who undertakes a particular course of action; even for him, "cost" evaporates at the moment a choice occurs.[66] Beyond this, because of the pervasive uncertainty that permeates all human affairs, these "forgone opportunities" cannot be objectively evaluated by the individual when choices are made. If this view is accepted, then no outside observer can ever know the "cost" of any action, neither estimated *ex ante* nor abserved *ex post*. All aggregation of money prices performed in the typical cost-benefit study becomes, on this view, irrelevant. These calculations simply do not represent the "costs" associated with the undertaking of a project, in either actual terms of, for instance, Adam Smith's "toil and trouble" or the modern Neoclassical formulation of "opportunities forgone."[67]

The Neoclassical reply to this line of argument is to invoke conditions of perfect competition as an approximation for true opportunity costs. To the extent that competitive markets utilize resources efficiently, the Neoclassical economist argues that they also approximate the actual opportunity costs associated with the use of those resources—and, further, that this "cost" is completely reflected by the market money prices that prevail in those markets or by the adjustments to those prices that analysts can make for the divergence of actual market data from the perfectly competitive model.[68]

To this rejoinder, Austrians reply that so long as actual markets remain imperfectly competitive, this method remains unsatisfactory. Beyond this, Austrians question whether or not the "competition" in perfect competition theory really has much to do with the real-world competitive process we actually observe. In that physical world, where time, uncertainty, and dynamic disequilibrating processes dominate all ongoing economic activity, "competition" is the day-to-day functioning of millions of separate (though often interrelated) markets. For example, in the physical world all firms face downward-sloped demand functions for their output. To say that these firms, no matter how contested their markets, are in some sense "monopolistic" and that they are exercising "monopoly power" is to measure actual reality by a decision standard that is too far removed to be of much use as a policy criterion.[69]

The Neoclassically trained cost-benefit analyst must carry out all calculations within the comparative-statics that constitute the body of microeconomic theory. When the static framework is utilized, it makes a good deal of sense to speak of "economic scarcity, " "constant demand/supply relationships," "constancy of consumer indifference mappings," and other techniques that are, at root, essentially thought-experiments where *ceteris paribus* conditions (such as the "state of technology") can be maintained outside a laboratory setting.[70] In fact, actual markets always work under dynamic conditions where little is fixed and information is always both scarce and imperfect.

All these considerations are important for any CBA, because the analysts would like their final cost number to have some reasonable resemblance to the actual cost of undertaking the project. Yet, at every step of this definitional process, objections and competing views can be profitably entertained by theorists. Suggested theoretical adjustments to current techniques in the light of the alleged subjectivity of cost are outlined later in this chapter.

Measuring Benefits

The same sorts of definitional and practical calculational problems that beset the attempts of cost-benefit analysts when they attempt to measure "costs" also impinge when they attempt to measure "benefits." The Neoclassical theory of consumer behavior disallows interpersonal comparisons of utility states. Since they are, by assumption, not observable, no cardinal measurement of their magnitude has been generally accepted in economic theory or practice.[71] Generally, Austrians tend to be in agreement with Neoclassical theorists that while utilities can be ordinally ranked, they cannot be cardinally measured.[72] Once again, Neoclassical theorists use money prices in trades as a proxy for value—but values and prices paid, whether observed or verbally stated, are simply not the same thing and should never be assumed to be identical. In simple terms, it has to be the case that individuals engaged in voluntary exchange value the thing they are attempting to acquire *more* than the thing they are surrendering to the other trader. It makes no difference that this surrendered thing is money itself. That is why more money prices do not accurately reflect "value" and why, for example, the Gross National Product does not (in spite of the contentions of some textbooks) measure the "value" of total output. Even the marginal consumer in a market must receive some consumer surplus, no matter how small, or that trade would not take place. The question of the *magnitude* of these surpluses remains conceptually understood, but it cannot be measured. Neither observation, introspection, nor questionnaires can lead us to an accurate number.[73]

Currently, the edifice upon which standard consumer theory rests contains several separate pillars. These different approaches can be termed verbal preference-based or action preference-based. Under the action preference-based category, there are two distinct (though related) approaches: revealed preference and demonstrated preference.[74] The verbal preference model employs the use of questionnaires to elicit preference patterns from consumers and to construct models of indifference curve mappings.[75] The revealed and demonstrated preference approaches attempt to model consumer choice by actual observation; this solves the issue of potential consumer lying but raises the issue of the extremely costly operationalization of this logistically difficult approach.[76] A further distinction is that revealed preference assumes a constancy of consumer preferences that the demonstrated preference approach denies.[77]

From the standpoint of operationalizing CBA, it matters a great deal which method the analyst chooses. Most cost-benefit analysts probably have not given this question much thought. They are willing to add together the money prices of observed trades, questionnaire trades, and even "shadow trades" as if they were always adding exactly the same kind of entities. But economic theory has generated more than one approach to this issue precisely because it *does* matter which foundational pillar one chooses to use as the basis for further theoretical development. A questionnaire that asked people how much they would value a new park in their neighborhood is a method for benefit "measurement" that is entirely acceptable to CBA, but it would be questioned by those economists who believe in the theory of demonstrated preferences. Could we trust the responses to such a questionnaire? Wouldn't people have a tendency either to overreveal their demand (if they thought it would help get the park built) or underreveal it (if they thought it would keep their taxes lower)? Do numbers added together in the absence of actual trades have any relevance to value at all? Austrians would answer in the negative, their guiding preference-finder being: "Watch what people do; don't listen to what they say." To which the Neoclassical theorist can reply: "What people say is, within certain parameters that I believe can be controlled for, just as accurate an indicator of their true values as what they actually trade for." Even beyond this, the Neoclassical theorists charge the Austrians with circular reasoning and tautological uselessness in the acceptance of demonstrated preference. Critics also add the contention that demonstrated preference implies cardinality for the measurement of utility.[78] These different ways of looking at consumer valuations and, hence, at the benefit issue are mutually incompatible. This dispute cannot be pursued further, but Rothbard[79] sums up nicely:

> Demonstrated preference...eliminates hypothetical imaginings about individual value scales. Welfare economics has, until now, always considered values as hypothetical "social states." But demonstrated preference only treats values as revealed through chosen actions.

When the CBA enters "benefit(s)" into the calculation, it has to be remembered that the numbers are either historical money prices or assumed valutions. If they are historical money prices, then two sources of potential error are the assumed constancy of preference over time and the inability to equate value with prices actually paid. If they are assumed valuations, it is not clear what meaning such numbers can ever have.[80]

The Changing Universe and Paradigmatic Commensurability

The ever-changing nature of social processes poses large—and, for the most part, unresolvable—problems for policymakers. When is a cost-benefit study to be done? How does the knowledge that the project has been proposed and is being studied affect the current preferences of those whose values are to be studied? (Or does it?) What changes start to occur after the policy is approved *because* of that approval? Can follow-up studies terminate the program? If so, why bother doing the original study? If not, how does the analyst allow for the changes that the passage of time creates?[81]

It is not a valid criticism of CBA to assert that many analysts who use such procedures are incompetent to do so or have hidden financial/political agendas (although that is sometimes the case). The individual analyst ideally has no control over the disposition of the final vote on the project except for the influence of the study. Certainly, every outside consultant/analyst knows what the funder wants to hear; not all such studies tell that story, however, which suggests that objectivity is possible for the cost-benefit analyst even if not (in theory) for the technique itself. In this regard, the questions raised in this section might lead to an extension of these techniques rather than to their abandonment altogether. Until CBA incorporates all the theoretical tools that are available, such studies will continue to be met with skepticism on the part on the public and those theorists and practitioners who understand where its weaknesses are concentrated.

There is no uncertainty about whether or not CBA is going to continue to be used in the future; it will be so used. Rather, the question is this: are policymakers willing to incorporate criticisms into the framework through which they will filter proposals for spending so much of other people's money? If the answer is in the affirmative, then this critical appraisal suggests the need for reevaluation of the following key areas of assumption and theory:

(1) Perfect competition theory as an empirical benchmark

(2) Assumptions concerning the value of taxed capital

(3) Assumptions concerning the social discount rate

(4) Assumptions concerning the nature of risk itself

(5) Assumptions concerning current definitions of cost and benefit

First, instead of the assumption that the actual world is somehow deviant from the model of perfect competition, we need a better theory of the competitive process than we currently have.[82] If we assume that the dynamic competitive process accurately demonstrates people's economic preferences by giving us some indication of their valuations of goods and services, then we are led to conclude that the "costs" of transferring resources from private use to public use are higher than assumed by

Neoclassical theory. It can no longer be a question of transferring "non-perfectly competitive" resources from the private sector, where they must exhibit "welfare losses," into the public sector, where they can be used efficiently to correct for "market failures" and "free-rider problems."

Further, as analysts, we must assume that market interventions have distortive effects; hence, the "costs" associated with resource transfers become unknowable, even within the Neoclassical framework. Adjustments to CBA are commonplace precisely because it has been widely recognized that the distortive effects of prior state actions must, as a first approximation, misallocate resources away from their most highly-valued uses and, second, alter the observed money prices in a pattern that would not have prevailed in the absence of such interventions. For example, the costs of legal services cannot be accurately reflected by the observed money prices in the market because of the massive state interventions into that profession that create artificially high salaries for attorneys—e.g., state certification of law schools and law board exams that artificially restrict supply, laws excluding non-certified attorneys from doing simple legal chores, and the writing of laws and regulations (mostly done by attorneys) that require expert legal advice before ordinary citizens can do much of anything.

No dollar estimate of legal "costs" can ever be accurate, given this network of existing interventions and misallocations fostered by prior interventions. How would a CBA study of the relative "efficiency" of the Legal Services Corporation versus private alternatives even make sense, when it must be based on such cross-sectional, distorted data bases? This problem remains insoluble even if the further complication of quality variations is set aside.

Except in those cases such as national defense, where the Neoclassical argument for public goods is strongest, arguments about "benefits" need rethinking. Every special-interest lobby argues for their spoils in terms of the benefits that are to flow to others. The usual practice is to locate a sympathetic analyst to perform a CBA that, indeed, "demonstrates" some sizable social cornucopia. Cost-benefit analysts need to begin to realize that, with the exception of the strongest possible public goods case (one is discussed in Chapter Four), the typical outcome of an objective cost-benefit analysis will be that net benefits of public transfers of wealth are negative. It is difficult to assess benefits within a model of voluntary exchange. How much more difficult, then, to attempt a quantitative measure of such benefits without even that model to fall back upon. Yet that is what CBA attempts to do.

The same sort of rationale applies to the issue of setting the correct rate of social discount with which to do a CBA. Different rates ought to be used, depending upon the policy that is being considered. The assumption of an unchanging rate over time needs to be abandoned. In public undertakings, as in private ones, a longer time-frame should result in a higher discount rate. Concomitantly, the purer the public goods case for the project, the lower the rate should be, but with the following potential exceptions:

a. some adjustment for the fact that capital might have been supplied involuntarily, and

b. people are not subjectively negative about the project to such an extent that the social rate of discount has to be adjusted for this fact.

Criterion (a) is necessary to adjust for the fact that capital that will finance many projects has been transferred involuntarily from individuals who (at least at the time of that transfer) valued it more highly for alternative uses. The Neoclassical contention concerning the "myopia" of the private sector simply proves too much. It is not an objective, scientific argument; it is rather a banner under which one can find virtually all types of extant government spending projects. Quite simply, there is no way to operationalize this concept so that there will be a "stopping point." For if the next generation's desires are to be divined today by analysts, what of the generation after that? Or the generations out to perpetuity? By what magic formula does CBA claim to divine these numbers? In fact, one could as well argue that it is the public sector that is typically myopic.[83]

Taking these criticisms and changes into the current CBA context, the following equation might emerge:

$$P.V. = \sum \frac{aB - vC}{(1 + qR)^t}$$

Where:

$P.V.$ = present value of the project under consideration

B = benefits of the proposed project

C = costs of the proposed project

R = social rate of discount

Note that a, v, and q are scalars: a is some rate change in dollar benefits because dollars are valued less highly than what they are traded for; v is an adjustment to costs that raises them (see the discussion above); q is a scalar that raises the rate of discount depending on the length of time the project will take to complete (and, presumably, endure), the purity of the public goods case for the project, and the project's capital size.

How is the "purity" of the public goods argument to be evaluated? The rationale for public goods has several interrelated assumptions. Each can be quantified, and the project can be evaluated against those numeric outputs. A decision point can be set so that if the project meets the decision value, a presumption exists that it might indeed be worthy of a public undertaking. If not, then no CBA need be undertaken, as the public goods rationale simply is not present.

It might be argued by some that a consistent application of these changes would lower the possibility of any public project being able to show net benefits. This is true, but it reflects the objectifying of the CBA technique; this development should, therefore, be welcomed by those who do not wish to use CBA merely as a way to legitimize all their favorite projects. Rationales for policy can still be grounded on moral or political foundations, rather than on the quasi-scientific foundation that CBA allegedly provides—and that is probably a good thing. The mantle of Science ought not and need not be used for such purposes.[84]

Now that the theory and practice of risk assessment and cost-benefit technique have been explored and some of the problems associated with them have been listed and discussed, we can progress to the evaluation of an actual case in the next chapter. If we were allowed to invent some policy problem that, in theory, would be understandable and justifiable through the application of these two types of analysis, few hypothetical situations could match the great swine flu episode of 1976. But we ought not to be surprised to discover that experts failed to utilize their most powerful and useful tools when it came time to decide on a policy option—even in a public health situation where quick and decisive action could have been imperative. Even in retrospect, these techniques do not support the decisions made in 1976, although we now have much better data and a better perspective on events.

4

Scientific Policy Meets Reality: The Swine Flu Episode

A thousand probabilities do not make one fact.

old Italian proverb

Fate laughs at probabilities.

E. G. Bulwer-Lytton
Eugene Aram

The purpose of this chapter is to demonstrate that, contrary to expectations, neither risk assessment nor cost-benefit analysis proved useful to decision-makers faced with a potential health crisis in 1976. This outcome should be viewed as contrary to expectations because, *a priori*, the swine flu decision seems almost ready-made for the two techniques. First, a plausibly accurate calculation of the risk of the epidemic's occurrence ought to have been completed. Second, a further risk analysis of the probable outcomes for Americans of just such an epidemic ought to have been completed. Lastly, a CBA should have been done so that the net benefits accruing to a massive, federally funded inoculation program might have been demonstrated to the satisfaction of both policymakers and voters. In reality, although such things were "done," the form they took bears little resemblance to what might have been expected. *Even retrospectively*, it remains impossible to carry out those calculations accurately. How, then, was it supposed to have been accomplished at the time?

The essential point of the analysis that follows is *not* to second-guess decision-makers in 1976 from today's perspective of 20/20 hindsight. Rather, the point is to explore carefully the basis for their decision and ascertain whether the best possible use of existing techniques was made or whether any beneficial change could have been made in the manner in which the decision was carried forward and implemented. The swine

67

flu decision, when viewed in the light of the evidence that will be presented in this chapter, was *not* the disaster that many claimed, nor was it a particularly glorious moment for scientifically-based decision theory. It was just another participant in the seemingly endless parade of programs in public policy as daily exercises in "muddling through." Should the swine flu program have been implemented? What, actually, were the *ex ante* risks of another pandemic? Can those numbers generate net benefits in a CBA? How were risk assessment and CBA analyses used—if they were? Finally, what role (explicit or implicit) did the objectivity-subjectivity dispute play in the decision, and how might that have been changed? This episode will provide a useful empirical test of the power that is sometimes claimed for risk analyses and cost-benefit techniques.

Public policy surely predates all attempts to examine its premises and rationally evaluate its outcomes. For this reason, risk assessment and cost-benefit analysis are rather recent historical appendages to the actual practices that they were designed to analyze. It has always been possible, of course, to appeal on an *ad hoc* basis to some set of "benefits" that allegedly flow from an existing policy—and then to implicitly infer that such a policy is, therefore, obviously valuable and necessary. Indeed, attempts to attenuate ongoing public undertakings will always be met with a chorus of "special pleading" from policy beneficiaries.[1] The truly important question, at least from the perspective adopted in this examination, is: how well does actual policy mirror *a priori* assessments put forward at the time the policy was being debated before its adoption? If those assessments were accurate models of actual policy outcomes, those practices thereby receive some "scientific" support. If not, the usefulness or objectivity of those same practices (and perhaps of the analysts themselves) might be called into question. One possibility is that the analysts are not properly utilizing the proper technique(s).[2] Another possibility is that the techniques themselves are defective—if, by "defective," we mean that they do not accurately model the future that comes into existence after the policies are implemented.

This chapter examines a public program that was debated both on the basis of risk assessment and cost-benefit theory, although official use of the two techniques was not necessarily accomplished in the usual way. The point of the following analysis is not to second-guess the original decision but simply to compare the projections and predictions of the analysts with the "facts" as they emerged and as they are now perceived. Further, it is not the author's position that this policy ought not to have been implemented because it "failed," although that is currently the generally accepted opinion.[3]

Our quasi-Popperian working hypothesis is: risk assessment and cost-benefit techniques were valuable predictors of uncertain outcomes. Naturally, the word "valuable" does not lend itself to a neat statistical test. How "closely" the modeling mirrors fact is a matter of individual taste and judgment, not an objective test statistic lying within or outside a particular range of predicted outcomes. For that reason, our attempt to falsify the hypothesis outlined above will be rhetorical, using statistics only in an illustrative sense.

Scientific Public Policy: A New Secular Theology

Before the twentieth century, most enacted policy was put in motion because people had become convinced that the policy ought to have been implemented. In brief, public policy debates tended to be "normative."[5] Appeals were directed to the voters' sense of "right" and "wrong" and, in the case of self-serving legislation, directly to their economic interests. By the twentieth century, though, the scientific revolutions of the Enlightenment had spilled over into the endeavors of social science. From the Progressive Era onward, the prevailing view in the academies and among "learned" persons was that rational social policy was not just a dream but, with the correct application of proper methods, could become actual reality. No more would policy have to be debated in the darkness of received doctrine and prejudice. Now it could be argued in the light of scientifically-produced "fact." Sidney Hook summed up this idea nicely when he attempted to define modern political liberalism. It is, Hook wrote, "faith in intelligence."[6]

It is now beyond argument that Hook's "faith" is the operational method that guides policy judgments. Decisions between complex social policies can now be made, according to this view, by appealing to the operational risks within a cost-benefit framework. According to the policy rationalist, if the risks are "acceptable" and "benefits" are greater than "costs," what person can argue that the proposed policy ought not be done, and on what inductive basis? It will not work to say that the proposed policy is "wrong," "immoral," "unjust," and a "waste of time and effort." These arguments are "unscientific" and "value-laden" with the citizen's personal, irrational prejudice. Why are they "irrational"? Because they might have been accepted "without a proper evaluation of the empirical evidence" from his parents or some other *ex cathedra* source or perhaps, worse still, he might not even be able to articulate and defend his position of support or opposition for the policy proposal being debated.

Against all the mere prejudice of average citizens, we have the testimony, beliefs, and inductive methods of the "policy expert." In theory, these elite individuals have been trained to evaluate policy objectively by the application of scientific technique so as to judge it solely on "positive" grounds.

This method avoids all the useless analytic arguing that characterizes the discussions of laypersons.[7]

The story that follows is testimony to a recent monument to this newly accepted twentieth-century faith. It is not the only such monument, nor is it the largest by any measure. Yet it is very instructive, precisely because it should have provided one of the *strongest* demonstrations of the correctness of the tenets of that faith. Other cases are, of course, interesting in their own right, and policy disasters are not difficult to uncover. That is not the issue here. There can be little doubt that we often *intend* to accomplish rational policy, but intentions do not usually jibe with policy outcomes. Economics can be viewed as the discipline that explores the unintended consequences of human action, but even this discipline is an insufficient device with which to see the entire future course of human responses to policy choice. Outcomes are, therefore, often different from predictive theory and human intent.[8]

Lessons of The Swine Flu Episode

In order to understand the analysis of the swine flu situation, a factual chronology of the events between March 1976 and March 1977 that comprise this strange episode in federal/state medical intervention is in order. Readers should keep the propositions below in mind as they move through the chronology. They are the judgments of the two policy analysts who had the privilege (by federal grant) to second-guess their unfortunate subjects concerning this policy choice.[9] They concluded:

(i)   There was overconfidence by medical specialists in theories "spun" from "meager evidence."

(ii)   Conclusions were reached "fueled by conjunctions" with preexisting "personal agendas."

(iii) There was too much "zeal by health professionals" in an attempt to make their lay superiors "do the right thing."[10]

(iv)   There often was "premature commitment"—deciding more than had to be decided.

(v)   There were "failures to address uncertainties in such a way as to prepare for reconsiderations."

(vi)   There often was "insufficient questioning of scientific logic and implementation prospects."

(vii) Finally, there was "insensitivity to media relations and the long-term credibility of government institutions."

Chronology[11]

In January 1976 at Fort Dix, New Jersey, there was a large "outbreak" of "respiratory disease." This was not at all unusual, given the time of year and the conditions recruits endure there.[12] It might have passed unnoticed, except for a bet between two doctors about the nature of the disease. As a result of this bet, specimens were sent to the New Jersey State Department of Health (NJSDH) for examination. By late January, NJSDH had requested some throat washings from Fort Dix for further analysis. Before that request, a recruit left his sickbed to take a forced night march. He later died, and his throat washing was sent on to the NJSDH.

By this time, the NJSDH had a total of nineteen washings. By early February, the NJSDH had identified two distinct flu strains, A/Victoria and A Port Chalmers, in fourteen of the sample washings. They were "unsure" about another five samples. These samples were sent to the Centers for Disease Control (CDC) in Atlanta. By the end of the first week in February, the CDC confirmed that five of the samples were indeed A/Victoria. More arrived by February 12th, and the CDC confirmed that they had isolated four swine flu samples as distinct from other types of flu.

Meanwhile, several overlapping samples were also sent on to Mt. Sinai Hospital in New York City. They confirmed the presence of the swine flu virus. This was a startling discovery for health experts, because the last suspected swine flu outbreak had occurred during the "pandemic" of 1918; it had caused twenty million deaths all over the world, with one-half million occurring in the United States alone.[13] Beyond that, theories current at the time suggested a major outbreak of influenza in, or around, 1979 (see more below on the eleven-year recurrence hypothesis). This was predicted to be another outbreak of some flu strain against which the population had little natural resistance, as in the 1918 case. Was 1976 "close enough" to 1979 for this to be that predicted event? Was swine flu, which doctors believed had lain dormant in person-to-person contact for over half a century, about to erupt once more? If so, then any delay might have proved disastrous, and health officials had strong reasons for becoming alarmed by the Fort Dix samples.

On February 14, meetings were held at the CDC in Atlanta to discuss the need to develop "sufficient" vaccine and to keep the public in the dark for as long as possible. Two days later, the World Health Organization was informed. Mt. Sinai Hospital had already begun to grow a fast maturing recombinant for use in the vaccine. After notifying the state health departments, the CDC realized that it could no longer keep the public uninformed, so it called a press conference. Although the CDC did not mention the 1918 pandemic during the prepared statement at this conference, the issue was raised during the question-and-answer period; naturally enough, it was the 1918 outbreak that became the focus of media attention in reports of the conference.

Private drug labs had already begun to experience difficulty in growing the virus for the development of the vaccine, especially the giant Parke-Davis. By February 22, the CDC had been informed of other isolated flu cases in Virginia, Pennsylvania, and Michigan; it also had knowledge of a previous Minnesota incident in its possession. By March 1, the Army estimated that "as many as five hundred" recruits "may have been" exposed to swine flu at Fort Dix. That news prompted a meeting of the Armed Forces Epidemiological Board at Walter Reed Hospital outside Washington, D. C. They recommended development of a trivalent vaccine—one that might theoretically be of use against the three strains—for the B Hong Kong, A/Victoria, and Swine Flu.

By mid-March, the CDC had become convinced that a major pandemic was a distinct possibility during the upcoming 1976-77 flu season. CDC decided to ask Congress for a $134 million program to vaccinate virtually every person within the United States. By the third week in March, many leading scientists had met with President Ford to discuss the available medical options, and Ford had discussed the funding situation with the Office of Management and Budget (OMB). On March 24, 1976, Ford decided to ask Congress for the money that the CDC had requested.

The end of that month produced an unanticipated problem. The private drug companies did not want to make the vaccine unless they were statutorily protected from liability from torts. Their insurers were discussing the possibility of a cancellation of insurance coverage unless this immunity was forthcoming. By the beginning of the second week in April, the House had approved the "Swine Flu Bill," PL 94-266. Insurance industry trade groups were still lobbying hard for tort exemptions for the drug companies. Their position was that inadequate frequency data made calculation of actuarially correct premiums impossible. A subsidiary argument was that insurers would only have a brief time in which to collect the premiums that were eventually levied.

On April 15, President Ford signed the bill. By the following week, HEW was searching for three hundred volunteers for a run of the test vaccine. HEW's in-house "media monitor" program showed that the swine flu program had the approval of 88% of the nation's editorial pages. Yet, on May 1, the drug companies were notified that their coverage was about to be cancelled.

Mid-May found the CDC requesting the state health departments to draw up "informed consent documents" for the implementation of the vaccine program. The sample form for vaccinees to sign that was finally agreed upon by all parties did not mention the potential for certain extreme side effects (including death) for those receiving the vaccination.

Disagreement among doctors about the severity of the problem and the correct federal response to it then surfaced in the media. Dr. Albert Sabin wrote an article for the *New York Times* in which he castigated the rush to vaccinate everyone and urged that vaccine be stockpiled for "high risk" groups in case it was later needed. Nonetheless, HEW prepared legislation to indemnify vaccine makers against all potential claims other than "negligent manufacture."

By the end of May, the dissenting voices had made an impact on public opinion which, according to HEW's in-house "media monitor," had fallen to "66% favorable." By then, insurance companies were threatening to remove all coverage from the vaccine makers. The Administration agreed with HEW and asked the Congress to indemnify the drug companies. Congress was not enthusiastic about releasing drug insurers from liability claims relating to the manufacture and use of the trivalent vaccine.

By mid-June, the CDC had developed "risk/benefit" statements for inclusion on the "informed consent" forms and had passed them along to state health departments. The risk/benefit statement claimed that minor side effects might accompany the vaccine shot, but they failed to alert the public to any serious potential side effects other than a possible case of "mild" flu. This, of course, might have been serious for older people in the high-risk groups, since flu can lead to fatal complications.

July brought the news that Congress planned to reject the insurance indemnification proposal. The Administration and the drug companies then made a deal on the issue, only to find that the insurers would not go along with it. Meanwhile, as it prepared what was perhaps the largest vaccination program in history, the government (through the CDC) began searching the world for evidence that swine flu was indeed the threat that the program's scope suggested it was about to become. No confirmed cases could be found.

By the third week in July, the insurance companies were sending detailed letters to HEW and congressional committee chairmen that listed additional reasons for their refusal to insure the drug companies during the implementation of the vaccination program. Besides those already listed above, they desired that the federal government shoulder all blame should the program "go wrong." Insurers had taken large (multi-billion dollar) losses worldwide for the last three years. They were convinced that liability insurance was never again going to be the money-maker that it had been before 1973.

In early August the program was given a boost when, ironically, the government misdiagnosed Legionnaire's Disease as swine flu. The government stated that either Legionnaire's Disease was swine flu, or that swine flu might cause Legionnaire's Disease. Both these fallacious statements were widely disseminated by the media. These misstatements were quickly corrected, however, but Congress was still stalling on the indemnification issue. Finally, after much lobbying and trading among the government, the drug companies, and their insurers, Ford finally signed an amended bill, PL 94 - 380. To gauge public reaction, HEW commissioned Opinion Research Corporation to poll the American public. ORC reported that 93% of those polled had heard of the program and that 53% stated that "they planned to get vaccinated."[14]

By September 3rd, the CDC and HEW had approved the vaccine for both distribution and use. Preliminary tests had turned up no serious side effects. By the tenth of the month, the liability situation had been completely resolved. September 22nd saw the giant drug maker Merck ship the first vaccine to the states. The plan was for almost 200 million doses to be administered by the end of December, even though the government had not yet decided what doses should be given to the young, the old, or those in other "high risk" categories.

The first swine flu vaccines were administered to the general public on October 1, 1976 at the Indianapolis State Fair. By the second week of that month, three senior citizens who had each received vaccinations at the same time at a Pittsburgh nursing home were dead. The following day, October 12, officials in Pittsburgh closed down the Allegheny County vaccination program until an investigation into those deaths could be completed. Nine other states suspended the use of any vaccine from the same batch that was linked to the nursing home deaths.

In a rush to judgment, the CDC denied, before any investigation could have been completed, that there was any link between the deaths and the vaccine. Indeed, the immediate autopsy results listed "heart attack" as the cause of death for all three individuals. The Allegheny County coroner, however, was skeptical of the "probabilities" of these deaths and took his fears public. By October 13th, fourteen people had died after having taken the vaccine. The CDC announced that this number was "well within the expected range."[15]

By the following day, the "body count"—as the media had begun to call it—stood at thirty-three. President Ford sensed that public support was slipping, but still trusted the health bureaucrats at HEW and the CDC; he was vaccinated, along with his entire family, on television.[16] That same day, Assistant Secretary of Health Theodore Cooper held a televised press conference during which he decried the "body count mentality" of the press. All states now resumed the vaccination program. On his national radio program, Walter Cronkite chastised his media colleagues just for having covered the Pittsburgh deaths!

By October 22nd, the "body count" stood at 41. No one admitted any connection between the deaths and the vaccine. By the first week of November, Sabin had written (in the *New York Times,* once again) that the "scare tactics" used to get people vaccinated were wrong regardless of the "intent" of the health planners, and that the vaccine should have been stockpiled as he had suggested in his earlier *Times* editorial.

The same week, with only one percent of the population vaccinated, the government sent urgent appeals to "inner-city radio stations" in an effort to persuade the "minority community" to go and receive vaccinations. A week later, the first confirmed case of Guillian-Barre syndrome was linked to a Minnesota vaccination. This was a potential side effect of inoculation for flu that could be fatal or paralyzing. Meanwhile, the state of Missouri finally confirmed a swine flu case, the first such confirmation in months.

By the third week in November, a dispute had erupted between a large drug company and vaccine maker, Parke-Davis, and the CDC over responsibility for two million "incorrect doses." Parke-Davis demanded to be paid for the doses, claiming that the CDC had supplied it with incorrect viruses from which to make that batch of vaccine. The government played down the problems with the program while it played up the one Missouri flu case. This tactic increased, for a time, the rate of vaccination across the nation.

These increases were short-lived; with the program's acceleration came more cases of Guillian-Barre syndrome. By early December, there were six known cases; a week later, several more cases had been reported. Eleven states had reported Guillian-Barre cases to the federal health authorities. By the end of the second week in December, a new consensus was beginning to form among the nation's "health experts" that the program ought to be shelved until the Guillian-Barre situation could be studied. By then, there were over fifty confirmed cases. Within a week, Undersecretary Cooper's resignation had been accepted by President Ford. Claims under PL 94 - 380 were approaching millions of dollars. The new head of HEW, Joseph Califano, demanded and received the resignation of the director of the CDC. Califano then proceeded to quietly (and permanently) close down the swine flu program.

Critical Analysis: Some Questions, Some Tentative Answers

This episode should have been a proud demonstration of scientific public policy in its most rational and effective form. First, there was the objectively determined health risk. In theory, both the probability and the effects of flu occurrence could be known because of the proven technique of vaccination and the received history of its side effects. Second, there was no need for endless normative debate about whether the public health arm of the government ought to implement such a policy, should it become convinced that it was necessary. This issue had been settled for decades to the satisfaction of the average citizen. In fact, no one seriously criticized this program on the grounds that the federal government ought not to have been involved with this type of undertaking in the first place.[17]

Finally, the material for a competent cost-benefit study was readily available, and the entire undertaking had both the blessing of current attitudes towards "risk" and a positive cost-benefit profile.[18] The reality of the actual outcome should give us pause, however, as we retrospectively examine this particular use of the vast federal policy machine. It might be argued (especially by those who like CBAs) that if better information had been available or if there had been more time to decide and gather extant information, then the policy might not have been carried out. But that, of course, is simply perfect hindsight. In fact, if policy could be formulated with perfect information, all such failures would be avoided. The problem is precisely that policy is *never* formulated under such conditions.

Several questions should be asked about this policy episode and its attendant effects:

I.   Why did HEW and its experts immediately decide, based on narrow evidence, that another epidemic was upon America and that universal vaccination was the only option?

II.   Why were the drug companies released from liability if the "risks" were so small?

III. Why was disengagement so difficult when the program's consequences began to materialize?

IV.   Who *ought* to have been liable for this policy?

V.   Did the policy analysts fail their methods, or did the methods fail the analysts?

VI.   Could it be the case that the CDC was correct after all, and that the actual outcomes were within "normal expectations?"

VII. Was the policy a "failure" in that the "costs" outweighed the "benefits?"

Conjectures and Retrospective Appraisal

I. Why did the CDC react so strongly to a few throat washings from Fort Dix, and why did they leap to the conclusion that mass immunization was the "solution" for a problem that had not even materialized? To answer this, one has to understand some of the fundamental axioms and corollary beliefs in influenza "theory" that were fashionable in 1976. Beyond that, one also has to understand the nature of large bureaucracies and the incentives of their individual members as well as the incentives of their top managers.[19]

The head of the CDC at that time was David Sencer, and he had run his organization for a decade. He "knew every facet" of his empire.[20] He was an enthusiastic proponent of "preventive medicine"; mass inoculatory programs are a major aspect of this approach when outbreaks of particular diseases seem imminent. The early reports that Sencer received from New Jersey, coupled with the advice he received from his own "experts" in the influenza field (perhaps colored by their perceptions of what they might have thought that he wanted to hear), certainly gave him an intellectual ("scientific") basis for a large-scale program. Some have speculated that this entire project was primarily driven by bureaucratic "empire-building" on Sencer's part.[21] Saving the United States from another killer pandemic by the decisive, prescient actions of the CDC—it must have been a pleasant vision to Sencer. While we cannot enter his mind, the possibility remains that these ideas may have played a role in Sencer's decision to act vastly and immediately. Perhaps some future study will question every major participant in the decision and attempt that way to determine what truly motivated Sencer. Such a procedure, however, will always be open to the charge that the principals' memories have faded or that their desire or ability to tell the truth is in question.

We can evaluate the entire incident introspectively.[22] What courses of action might average persons have taken as more information became available? Their initial reaction probably would have been to seek confirming evidence that swine flu was indeed about to break out. Had any other cases been reported anywhere in the world? Some persons, while waiting for confirming information, might have quarantined Fort Dix. If convinced that the next flu season would bring additional cases of swine flu, then they might have appealed to the drug companies to begin making a vaccine that would be available to doctors and health officials should the need arise—perhaps even suggesting that tax receipts be used for costs. A few people, no doubt, would even have opted for a *laissez-faire* alternative.

Notice that David Sencer did *none* of these things. The option that he immediately lobbied the Administration for was total inoculation—at a projected federal outlay of $135 million. That lobbying was almost certain to prevail, once the possibility of alarming the public became a virtual certainty.[23] Sencer, quite naturally, couched his memo to Ford in scientific (that is, positive) terms, but in retrospect his actions seem to fit the

Buchanan-Tullock public sector "rent seeking" models.[24] This public choice model suggests that public employees, especially the heads of bureaucratic departments or agencies, seek to maximize their own benefits, which usually takes form as the expansion of their personal empires within the government.

Sencer's request for mass action within a very short time-frame was supportable by appeals to a particular theory of influenza occurrence and spread which was then in vogue.[25] One-fifth theory and four-fifths empirical correlations, this view of influenza epidemics used historical evidence to predict a great worldwide breakout of a new (or long-dormant) influenza strain in or about 1978-79. Was it occuring early? If indeed it was a return of 1918's swine flu strain, then it might be real trouble, since no one below the age of sixty would have had any immunity to the disease.[26] Was this actually going to happen, as the theory predicted? The CDC applied probability theory at this point; considering the number of scientific experts it could have called upon for answers, it surprisingly made a great mess of the entire endeavor.

Many officials involved with the decision to go forward with the inoculation program gave after-the-fact justifications. Many uttered the same haunting line, as if they had all memorized it for future use: "The probabilty of another 1918 was *positive*." For example, there is the 1978 interview conducted by Neustadt and Fineberg with HEW's Undersecretary David Mathews, during which he stated:

> As for the possibility of another 1918...one had to assume a probability *greater than zero.*[27]

Such a contention is trivially true; it is therefore an insubstantial foundation for policy undertakings. Further, the probability estimates available to the decision-makers were, they later admitted, purely subjective in nature. As Neustadt and Fineberg relate it:

> Most (CDC experts) seem to have thought privately of likelihoods within a range from two to twenty percent;...these probabilities...were based on personal judgement, not scientific fact.[28]

Silverstein also describes a meeting in which misuse of the probability concepts arose:

> Mathews questioned Sencer and Meyer closely on two issues: The probability of a pandemic and the possibility that enough safe vaccine could be manufactured. *Neither could put a number on the probability,* but both agreed that it was *greater than zero.*[29]

Several implications emerge from these quotations and their context. First, Neustadt and Fineberg are no more sophisticated in their understanding of probability than the original experts. They seem to believe that there ought to have been a difference between the judgments of experts and "scientific fact." Their evidence suggests, though, that there was no such distinction. Mathews seemed to believe that he was asking Sencer and Meyer about two definitionally different realms when he requested the *probability* of pandemic while only asking for the *possibility* of vaccine production. In fact, he was asking for the same analysis: namely, the subjective judgment of "experts" on both issues. Beyond this, at least two facts are also suggestive. No one had a "number" to give to Mathews or anyone else, because no such number could have been calculated with any precise meaning. Also, since all real-world events have a probability "greater than zero," nothing important was actually being said when this statement was applied to the possible occurrence of another pandemic. Nevertheless, by the time the proposal was forwarded from the CDC to the Administration for action, these "probability calculations," *subjectively obtained but objectively used*, had become blown completely out of proportion. After all, a two percent positive chance of occurrence is fifty-to-one *against*. Yet on this basis (and on extremely pessimistic assumptions about the ability of the population to do better this time around than they had sixty years earlier under much more primitive medical conditions), the government acted precipitously and—in the perfect vision of retrospective hindsight—needlessly.

Since the program was not based on a foundation of accurate risk calculations, what gave it its early force? Speculation is inevitable here; Sencer's role has been criticized in detail, as have the political constraints that operated on the administration. Each person in the chain of command played a role, but the CDC definitely was the catalyst. Neustadt and Fineberg write:

> [In recommending the $135 million mass inoculation program to Ford] Sencer, in his own mind, may have been playing the hero; if so, he was but the first who did. Mathews, Cooper, and Ford...would follow.[30]

What is not in doubt is Sencer's hysterical presentation—based solely on his incentives and the subjective estimates of his staff—to Ford. Ford, perhaps commendably, reacted decisively even though both he and his aides realized that it was he who would take the blame if the program went awry. What needed to be emphasized then (and has still escaped attention) was that the experts were operating solely on grounds of subjective probability as well as that the outcome they subjectively predicted *a priori* to be very likely was that a pandemic was *very unlikely*. Thus, the enlightened guesses of the consultants and influenza experts were quite accurate. There was to be no 1976 pandemic. Unfortunately, what *was* to be turned out even worse for many innocent individuals.

II. Why were drug companies granted immunity from tort liability in light of the prevalent expert opinion that side effects were of trivial concern? The answer lies in part with Congress. Having created media and public pressure by the CDC's overplaying of the possibility of another 1918 epidemic, Congress found itself in a delicate position. Once a hastily conceived program starts to roll forward, and the press and public become convinced that that program is the solution to problem *x,* it is difficult to decelerate spiraling expectations as politically interested groups begin lobbying for its implementation. After all, Congress was debating this issue after drug companies had already begun—at the government's urgent request—to make the vaccine. To stop after that and deny the public the already-created vaccine would have been a political decision no one in Congress or the Administration would likely have made. Congressmen are not medical experts; in order to decide such an issue, they rely on the expert testimony of the same persons who dream up programs like this. Their decision therefore seems very understandable. If they delayed or killed this program, the results would be politically catastrophic—if, as some experts had predicted (as they were told), swine flu was coming. Their constituents would want to know why no vaccine was available; their votes in the next election would hardly be in doubt. There was also the seductive appeal of the companies and their insurers: just make the *government* liable. This, of course, translated into a plea to make the flu victims liable as a taxpaying group. But, after all, weren't they going to receive the benefits of the this policy in the first place? Besides, everyone was used to thinking about federal expenditures as the government's money anyway.

Congress, to its credit, was reluctant to grant indemnification to the drug makers outright—but, to its discredit, its reasoning was incorrect. Congress wanted the companies to be liable because many congressmen simply did not like drug makers or their insurance carriers. They thought the companies were putting something over on them by their appeals for indemnification, and they were not about to give the companies what they thought the companies wanted. The better attitude might have been to inform the companies that either they would be liable and be allowed to make a profit on the vaccine, or they would be protected from torts but be compelled to supply the vaccine at cost. The companies might have argued that an urgent request from the federal government to do a particular thing is hardly a matter of voluntary action. They also could have argued that with a federal request made within a remarkable short time period, they ought to have been held blameless.

In retrospect, the insurance companies, regardless of their standing with Congress or anyone else, proved to be correct in their assessment of the risks involved in mass inoculation while the government experts consistently understated those risks in order to get the program moving quickly. The more accurate position of the insurance companies was no doubt based on their natural inclinations to proceed conservatively, especially where risks are not calculable. If the companies knew that, why was that viewpoint (made available to Congress and the Administration during the controversy over continued coverage) dismissed by the government as incorrect? No simple answer can be given, but the ubiquitous notion of "political considerations" has great appeal as a possible explanatory variable. In a sense, Congress and the Administration were only pursuing a course of action that many economists should have applauded, since certain welfare theorists had argued that governmental programs were riskless, or at least less risky than alternatives in the private sector.[31] If the insurance companies refused to insure the program, that was only because they faced private risk, whereas the state could act because it could socialize that risk across the entire universe of taxpayers. When theory and reality clash, as they often have, it usually leaves cemeteries more crowded than before.

III. The government's inability to disengage from the inoculation program as one difficulty after another plagued it—up to and including the deaths of participating citizens—is all too typical of large, governmentally-run enterprises that seem to take on a life of their own as major interest groups become active. Once a policy is started—such as NASA and the space shuttle, or the Strategic Defense Initiative, or the Energy Department's "synfuels" program—the closing down of such an undertaking requires:

(1) compelling reasons, and

(2) public awareness of those reasons with concomitant political pressure forcing Congress to stop the program(s).

The presence of compelling reasons is not by itself taken in politics to be sufficient grounds for terminating any ongoing program. The slow pace and insufficient breadth of the closing of existing military bases provides a perfect example of this type of argument. The case of the base closings demonstrates that, no matter how much people want to cut military spending, no matter how obvious the methods are to do it, there is always resistance from entrenched economic interests. Further, since these projects are not run on a profit/loss basis, it is difficult to demonstrate *compelling* reasons.[32] If the government had been successful in preventing the public from finding out about the "body count," the program—although some insiders might have been convinced that reasons to end the program were compelling—could have gone on a good deal longer. Had the CDC actually believed its own press release about the "expected death count," they would not have had any reason to stop the program on a purely cost/benefit basis. This is true because of the nature of probability analysis and the tendency of virtually all inductive probabilities to be positive.

IV. Who *ought* to have been liable for this program's unfortunate consequences? The federal government instigated it, micro-managed it, and funded it (through taxpayer support, of course); it must be, then, the ultimately responsible entity. In retrospect, this seems a fair outcome, placing the blame where it belongs. Unfortunately, the monetary burden cannot be placed there as well, since the government has no funds of its own. Therefore, a perfectly just solution for situations like these will always remain elusive. It is true that the drug companies seemed to want something for nothing. In this case, they wanted the profits without the risks. Left alone, however, they never would have become involved in such an undertaking in the first place. The liability position of firms involved in manufacture for the federal government is not as clear as it ought to be, due to the government's vacillation on the issue of manufacturer liability for federal contract work. In such cases, no just solution to issues like these will ever be possible.

V. Did science and its techniques fail the medical experts, or did the experts fail science and its techniques? This issue goes to the heart of the arguments discussed thus far in this book. Advancing medical technology and medical opinion in 1976, appreciated and well-researched as it was, supplied few strong answers where influenza was at issue. Influenza was (and is today) a "slippery phenomenon...(N)ot much was known about pandemic spread."[33] It is therefore surprising that arguments based upon theories that had very little correlative evidence were so readily accepted by policymakers. Two theories of disease—that epidemics were separated by eleven-year intervals and the theory that, should influenza strike again, it would spread with "jet speed" and thereby frustrate all attempts to stockpile vaccine—were routinely taken to be true by the CDC, even though a good deal more research needed to be undertaken before such conclusions would warrant the degrees of belief demonstrated by policymakers at the time. In retrospect, the theory of "jet speed" itself seems to be a convenient rationale for arguing against Sabin's stockpiling advice.[34]

No theory of probability operates very well if it is required to produce an estimate for a unique event without either a class of "priors" (see

Chapter One) or empirical data with inferential frequencies. There simply was no legitimate estimate available for the probability of the return of swine flu, or the probability of its rate of infectious spread should it return. All such numbers used by analysts were subjective; they ought to have been expressed to their superiors *in those terms*. They were not. Subjectivity of estimation is not tantamount to being "untrue"—for that reason alone, the actual situation might still have been defensible (and, as detailed below, might remain defensible) on grounds of expert intuition. It is just that intuition is a poor substitute in this era of science worship for numerical estimates based on accepted scientific procedures.

In the fourteenth century, the Black Plague killed half of Europe. It then receded and has not recurred, despite the fact that humans are not immune to it and rats and fleas are still very much with us. What explains this dormancy? Why did it leave in the first place? We have only hunches about the answers. The data available in 1976 did seem suggestive of an increasing frequency of flu epidemics, but was that data reliable? In other words, was that an accurate picture of reality across the entire century, or better data collection, medical reporting, and tabulation? No one knew the answers to these questions in 1976. In fact, experts were not even certain that the 1918 pandemic was caused by the swine flu virus.[35]

The eleven-year hypothesis was summarized by its author and best known advocate, Kilbourne, as follows:

> . . . in every instance of a double antigenic shift (in both H and N antigens) *a new influenza pandemic had swept the world*. In modern times, such shifts seemed to be coming with increasing frequency, and recent experience seemed to suggest that a cycle of approximately eleven years had been established for their introduction.

> Major influenza strains were thought to recycle approximately every sixty to seventy years.[36]

Notice the characteristically skeptical voice of the scientist here: "seemed to be coming," "seemed to suggest," "were thought to recycle." If the eleven-year theory was true, then why the alarm in 1976—three years ahead of schedule? If time periods were shortening between the appearances of new viruses, then an alarm would have been justified, but the eleven-year theory was simply untenable. An additional problem for that theory was the even shakier evidence that purportedly supported the "sixty- to seventy-year" view of the reemergence of any particular flu strain. One or two observations do not a theory make, in spite of the fact that economists sometimes refer to the Long Wave "theory" of the Russian economist Kondratieff.[37]

Suppose that the decision-makers had placed a subjective estimate on the "truth" of the eleven-year hypothesis. What might that number have been? It is believed now that the hypothesis was false. Assume p = 0.70 (that is, a 70% probability) for an event A, where A is defined as "the eleven-year hypothesis is true." That leaves ample room for the possibility of earlier or later occurrence: namely, p = 0.30. Define B as "swine flu is coming again" and C as "the result is to be another 1918." Assign B a value of 0.20 (experts apparently did on a subjective basis),[38] and C one of .10. Viewed this way, virtually no one believed that anything as severe as 1918 was coming in 1976, even if it was the case that another swine flu outbreak was on the horizon.

The calculations, subjective or otherwise, cannot end at this point. Add to them the mutually exclusive events X and Y, where X is defined as the probability, however calculated, that "an effective swine flu vaccine can be developed within a particular time frame"—that time frame being, in the case at hand, "a few weeks." Event Y, then, is the probability "that such an amount of vaccine, even considering that it can be manufactured in the time frame, could be delivered through preexisting and/or new mechanisms in time to beat the spread of the disease." Suppose that we had assigned, given our strong modern faith in science, a large p = 0.60 to event X, and a somewhat more pessimistic p = .30 for event Y. Now, assume a last event Z, defined as "the vast majority of Americans will in fact voluntarily go to the mechanisms to receive these vaccinations." This event can be seen as depending upon the occurrence of events B, X, and Y. Assume an optimistic p = .50.

Perhaps a decision tree summary would be helpful in interpreting the possible probabilities (outcomes). Beginning in the upper left-hand corner, follow the path that leads through the series of events that culminate in the swine pandemic envisioned by the CDC. Each event is assigned a flu probability, and as one moves through the diagram, the mutually exclusive calculations (multiplications of each tree branch's number by the one before it) are finally applied (multiplied) by the relevant population. The relevant population is *not* the entire population of the country, because vaccine cannot prevent flu in those persons who do not receive the shots; no one expected the entire population to be vaccinated. The 150 million figure is itself generous in its estimation of population compliance.

We now have enough assumptions about influenza (even though we now question the theories widely believed in 1976) and enough of a subjective probability framework so that a preliminary calculation can actually be finished. Suppose that a person was carefully attempting to make the key decision that the CDC's Sencer made, but from within the theory framework explained and advocated in an earlier part of this book. What

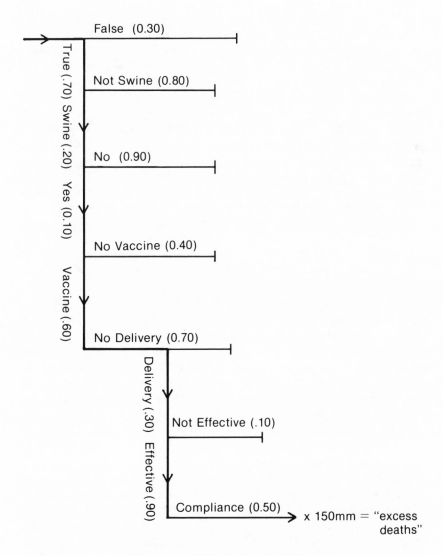

(Figure 4-1: Swine Flu Decision Tree)

might that decision have been? Let us begin by assigning a generous p = 0.70 to event Z whenever the *BXY* chain produces an outcome of 0.50 or greater. It is fifty-to-one *against,* according to this method, that a virulent outbreak of swine flu was about to occur. Given such an outbreak, there was still only an 18% chance that enough vaccine could be produced, delivered, and distributed. The mass inoculation option was therefore even then unsupportable *on the very grounds* that Sencer used while proceeding to implement it.

Of course, there is still the argument from "excess deaths"—the "if only one life is saved, it was all worth it" contention. Policymakers often deal with this sort of argument; it is usually made by persons who either have no conceptual knowledge concerning opportunity costs or persons who know but simply don't care for that line of justification. Going back into the model for a moment, assuming identical mortality rates with the 1918 pandemic (as some cost-benefit analysts were later to do—see the next section), then what was the number of expected "excess deaths" under this model? Expected deaths in excess of those predicted across the 1976-77 flu season would have been $(.70)x(.20)x(.10)x(.30)x(.60)x(.35)x(.50)$ times 150 million (see note 39)—approximately 56,700 excess deaths. To put this number in perspective, 1957's Asian flu outbreak was estimated to have caused 70,000 "excess deaths."[39]

There is no way, of course, to validate this estimate, nor to second-guess policymakers as to whether the expenditure of $135 million (in 1976 dollars) was a reasonable response to such a predicted event. But what is not at issue is that the analyses done in 1976 did not make the best use of probability theory or risk assessment technique, even insofar as they related as to whether or not people would get the shots. A rational case can be made, given the experts' own numbers, for having chosen *not* to get inoculated even if the epidemic had materialized.[40]

VI. Could it be the case that the CDC was correct after all, and that the subsequent events fell within "normal expectations"? It might be argued that the CDC, entrusted as it is with protecting the health of all Americans (or all persons within American borders) must always err on the side of caution; it must always, therefore, attempt to spare no expense in carrying out its mandate. That position degenerates rather quickly, however, into making the CDC into some kind of medically totalitarian entity. This is not acceptable within the American framework. The CDC, no less than any other public agency, is ultimately accountable to the people and therefore must justify its programs and expenditures to them.

Neither the CDC nor any other agency can ever be "refuted" on the issue of assigned probabilities, since all reality merely confirms, but never refutes, probability estimates (as we have seen in Chapter One). Trivially, any agency could then base its programs on whatever numbers it wished to attach to them, much as public utilities sometimes include questionable items in "costs" when they appear before rate-setting authorities. These disputes are never-ending and probably unresolvable. In this sense, the CDC was correct in its decision to proceed, since all probabilities are positive and another 1918 can always be just around the next "time corner." If all this seems unsatisfactory in some way, it is undoubtedly due to our residually strong belief in the scientific basis for public policy decisions. But *ex post* occurrences, while they are the mechanism by which we often judge public policy decisions, are (by the methodology of the policy apparatus itself) not sufficient to discredit those policies. To be fair, since risk assessment relies upon probability theory, we cannot condemn policy for either correct or incorrect forecasting of the future. For the future *always* (if properly understood) confirms classical probability theory.

The last weapon in the rationality arsenal is cost-benefit theory. Through the application of that technique, could we once and for all make an evaluation of the swine flu program that is scientific and objective? In fact, one such evaluation was undertaken;[41] it will be appraised in the following section of this chapter. After the analyses in Chapter Three, we ought not to be surprised that, here, too, absolute answers will continue to elude us.

The retrospective study of this policy seems to make some of the central arguments in this book clearer. First, all policy requires moral (that is, normative) justification if it is to win the continued support of the citizenry. Second, even if the reader disagrees with the foregoing proposition, since it invokes peoples' values rather than empirical fact, there simply is no operational induction-based alternative.

VII. Did the policy fail because the costs outweighed the benefits? Although the best time to perform a cost-benefit study is before policy is implemented, a cost-benefit study supporting the already-ongoing swine flu program was undertaken by Stephen C. Schoenebaum, Barbara J. McNeil, and Joel Kavet; the results of this study were published in *The New England Journal of Medicine* early in the fall of 1976.[42]

The authors began with a questionnaire that was designed to elicit from "influenza experts" the "probability" (purely in terms of subjective hunches) of another pandemic like the one that occurred in 1918. The survey data were followed by telephone contacts that provided "feedback" to the authors of the study from other groups of respondents. These telephone discussions ultimately resulted in a median variation of response of less than 10%, which is to say that the range of variation in the expert opinions solicited had been narrowed to workable limits. (Whether this conforms to the Delphi theoretical technique is not an issue here.[43])

The cost-benefit study defined "benefits" as those "costs" that would not have to be borne if the program were carried out under different sets of assumptions as to targeted groups, and so forth. Into their calculations were placed the direct, indirect, and "intangible" costs associated with flu distress. "Direct costs" are the *financial transfers* from influenza patients to the medical establishment, which was defined as "doctors, hospitals, and drug companies."[44] It needs to be pointed out here that Neoclassical cost-benefit analysts would not accept this definitional framework for aggregating actual opportunity costs. For reasons discussed below, this method has severe problems. "Indirect costs" would be the "forgone productivity" of the sick, as well as premature deaths from flu exposure. These numbers were estimated as follows:

(1) 1975 estimated daily earnings used as productivity proxy

(2) assumption of 2.8 "workdays lost" on average for those ill with the flu

(3) the present value of lifetime earnings for the prematurely dead would be discounted at nominal interest rates of 4, 6, and 8 percent

It was assumed that the total cost for the vaccine to the general population was 50¢ per dose, a total of $100 million. If only those over 24 years of age were to be vaccinated, then the cost would have been $60 million. If only the "high risk group" (defined as those over 64) were to be vaccinated, then the cost would have been $24 million.[45] (As discussed in footnote 26, however, the assignment of the elderly to the high-risk category contains an interesting anomaly.)

How many years of life would be "saved" by inoculating the populace? That would depend upon the following formula:

Cases averted = P(cases) x P(epidemic) x P(vaccination efficacy) x P(vaccine acceptance by public)

Years of life saved were calculated as follows:

Years saved = P(epidemic) x P(efficacy) x P(acceptance) x years lost

The authors assumed that a 1918-like epidemic would cause an additional fifty million flu cases. There would be 50,000 excess deaths, 38% of which would occur in individuals over 65. The median subjective probability estimate that was obtained from the questionnaire was 0.10. Having this number, the authors then attempted to calculate the other parts of the formulae. Assumptions were made about the compliance of the public with the swine flu program.

Group	Compliance Rate
5-24	75%
25-44	60%
45-64	60%
65 +	70%

Along with these compliance rates would come side effects; the authors estimated that there would be 307 excess deaths due to the actual administration of the vaccine. Based on these estimated figures, the authors calculated that the total "costs" of not vaccinating would amount to almost $6 billion.[46] Interestingly, the authors seem to accept as axiomatic that the 307 deaths that are due to the administering of the vaccine are an acceptable price to pay for the benefits alleged to accrue to the entire program.

Under their scenario, 1,130,000 years of life for the general population could be expected to be saved; 954,000 years of life for those 25 and older; and 857,000 years of life for those in "high risk groups." The authors admittedly ignored many "intangibles" in their calculations—for example, "pain, suffering, and grief associated with illness or vaccination." Days lost from school and the disruption of community services were also ignored. So was the indemnification/torts issue that plagued the program, on the grounds that it was not a significant factor.

Without splitting too many hairs, a careful analysis of this study can demonstrate the cloudy and uncertain nature of any cost-benefit study. The large quantity of estimated data is characteristic of such endeavors, because the actual data needed are almost never at hand. Notice that every single piece of data used by the study's authors is, in some sense, manufactured—and sometimes just invented.

Begin with the "probability assessment" of the likelihood of the epidemic's occurrence.[47] The authors were supplied with tiny samples of opinions from author-designated "experts," although there was a wide variation in response. In fact, one expert set the probability of a pandemic at 0.02 while another set it at 0.40—that is, 20 times more likely! The authors "massaged" a median response of 0.10 from these varying responses, but only from the smallest sample. Further, these "experts" were not always expert at the same thing. Which opinions were, therefore, "more expert"? They cannot all assumed to be equal, since their medical specialties varied.

The authors also made some assumptions about what constitute "costs" and "benefits" that are actually arguable, even though they are commonly used in a similar fashion by the authors of other studies. Much of what the authors chose to add together as "costs" were, in reality, simply "income transfers" between individuals and other individuals or institutions. If I do not spend a dollar on x, then I can spend it on y. The "cost" to me of x is, in strict Neoclassical opportunity cost analysis, precisely the forgone opportunity y. Yet there is no overall "social cost" if I choose y rather than x, even if x was my first choice. Nor is there any way of knowing precisely what the "cost" of any such action really is, since the actions are never taken. Consider this case: I go to the doctor because I have the flu; while I'm there, the doctor discovers something else that is wrong, catches it early, and is able to cure me. I would have died, had I not gone in—but I went in because I had the flu! How does one quantify the "cost" to me of that visit to the doctor? The authors assume that every dollar spent on doctor/hospital care that would be generated by flu symptoms can be added together as some kind of "social flu care cost." Up to a point, it is easy to agree, as long as we are clear about what we mean by "cost." But the number so obtained will not be an accurate reflection of the "cost" of a flu epidemic because of situations like the above, as well as the fact that, even though the flu causes grief, being out of school or off from work are things many people enjoy. There are also people who enjoy being sick because of the effect it has on their relationships with others. Against this, there is the output (whether educational or work-related) that is indeed forgone by the absences that might provide psychic net benefit to workers or students. Even here, however, it is a two-edged sword, for the lost "productivity" of non-medical workers who get sick is matched by a (presumed) reduction in the productivity of the medical sector that would have treated them. In fact, if everyone stayed healthy, the productivity of that massive sector of the economy would fall to zero! This confounding effect follows from the fact that one person's "cost" is another person's "income." These effects are sometimes referred to as "pecuniary externalities."[48]

Is it even clear-cut that "years of life saved" is an unambiguously defined "good" that would flow from the program? It depends, once again, on *whose point of view is being taken.* It is clear that a massive program of this sort, to the extent that it might be successful, would prevent the deaths of individuals whom other individuals might not have saved, who will in the future commit criminal acts, some of them quite brutal and fatal, and many who will die more horribly because they were saved by the program. We cannot ask them which they prefer, of course, and the entire argument might sound excessively "nitpicking" in nature, but these facts cannot be qualitatively added or subtracted nor even remotely estimated. One can only muse philosophically about whether a mass health program that spared a future Hitler or Pol Pot or Stalin could ever show "net benefits."[49]

The use of wage data or other productivity proxies is also a necessary evil, but it needs to be carefully handled. It is simply illegitimate to argue as follows: x number of cases will occur that would not have occurred, and y is "average productivity," therefore the social loss is xyz, where z is the "average number of days lost from work." The reason may not be immediately clear, but leave aside the issue of whether or not the productivity/wage information is accurate for a moment, and focus instead on the typical American worker's habits.

Workers typically allocate a certain number of days per year for "sickness"; companies do the same. If a person really is sick with the flu and misses 2.8 days of work, then that person will simply factor that into their total missed work equation for that year. There really is not "lost productivity" in the sense cost-benefit analysts claim because Parkinson's Law comes into play: work expands so as to fill the time available for its completion. If these workers didn't miss work because of the flu, they would miss roughly the same amount of time over the course of the year anyway, and their output wouldn't vary greatly (if at all) from what it would have been in the first place. Indeed, a severe enough pandemic that actually began affecting output would simply increase the pressure on the remaining workers to meet production demands, and their productivity would increase concomitantly to pick up the slack.

Support for this counterintuitive proposition can be gleaned from the *Historical Statistics of the United States*. If, in fact, influenza caused productivity to decline, we might expect that it, along with the GNP, would fall during flu years. In fact, this didn't happen in the swine flu years of 1917-19, or the 1956-58 period, or during the 1967-69 Hong Kong flu period. In fact, both productivity growth per worker and Gross National Product are positive throughout these interludes.[50]

1917	Output Per Man/Hour (in dollars)	GNP (in billions of dollars)
1917	33.3	60.4
1918	36.4	76.4
1919	38.1	84.0
1956	94.6	419.2
1957	97.2	441.1
1958	100.0	484.3
1967	131.3	793.9
1968	135.3	864.2
1969	136.4	930.3

Naturally, one could argue that the rates of growth would have been even higher in the absence of flu outbreaks. Counterfactuals, however, cannot be conclusively demonstrated; they can only be conjectures. The confounding problems associated with isolation of the effects due solely to the influenza and not to macroeconomic pressures are probably insurmountable. It is fair to state, however, that average growth rates do exhibit a downward trend between the years 1918-19 and 1957-58. Yet the years 1968-69 do not exhibit this feature. One possible explanation is that, as we move closer in time towards the present, the economy has become more static in terms of the behavior patterns of workers as they have been given more and more fringe benefits, such as a certain number of "sick days" and/or "personal days." Under this type of arrangement, the number of days missed from work will tend to expand to the number expected and allowed for by management. In any case, explaining past data is always mostly guesswork.

For purposes of computation, then, we have seen that cost-benefit analysis must rely on the "everyone is identical" hypothesis so that "costs" and "benefits" can be appropriately aggregated. In fact, it is impossible to add two "costs" together if one adheres strictly to an opportunity cost framework.[51] The assumption of homogeneous people is an unstated but absolutely necessary condition of such aggregates. If you and I both go to our family doctors because we have flu, and each of us surrenders a check for $75, the "costs" are *not* $150. That would mean that each of us gave up alternatives of equal value and that these values can be added. In fact, such addition requires the grouping together of apples and oranges, for we cannot suffer the same "cost" even if money outlays are identical.

There is the further problem of accounting for the "excess deaths" (estimated at 307) that would occur from the medicine being administered. Leave aside for a moment that a doctor's recommendation that a patient receive a shot that ends up killing him violates the Hippocratic oath's injunction "First, do no harm." How does the analyst value these 307 people? We are trapped in the old utilitarian box of suggesting that their lives can be sacrificed because their number is less than those who will be "saved" by the program. This argument is unpersuasive as well as invalid. It requires an accurate valuation of human lives, and that issue is far from resolved in torts, even if such "answers" are routinely given by our court system.[52]

One final point is alluded to by Schoenebaum, McNeil, and Kavet, but they fail to see the implications of it. They state:

> We suspect that most areas of the country do not have the flexible resources (personnel, materials, and sufficient supplies of vaccine) needed for immunization of the entire target group within two to four weeks of hearing of an outbreak elsewhere. Such rapidity is essential.[52]

The concept of opportunity cost infuses this statement, yet in their calculations there is no adjustment for the harm that this program will cause by draining resources from alternative uses, thus making personnel, laboratory facilities, and delivery systems less accessible to those suffering from other illnesses. Such an estimate ought at least to have been made and subtracted from their "net benefit" calculation. Once again, however, they would have been simply inventing numbers, for how can anyone know the "answer" to this objection?

All things considered, it is difficult not to agree with the judgment of *New York Times* editorialist Harry Schwartz:

> The sorry debacle of the swine flu vaccine program provides a fitting end point to the misunderstandings and misconceptions that have marked government approaches to health care during the past eight years....
>
> Any reasonable effort to assign responsibility for this state of affairs must call attention to at least the following elements....
>
> (1) The scarcity in the White House and in Congress of officials with sufficient sophistication in medical matters.
>
> (2) The excessive confidence of the government's medical bureaucracy and its outside experts in urging the vaccination program on the country while playing down the uncertainties arising from the fact that medical science still knows comparatively little about the origin and spread of influenza....
>
> (3) The self-interest of the government health bureaucracy which saw in the swine flu threat the ideal chance to impress the nation.[53]

Or finally, from Neustadt and Fineberg:

> What a basis on which to build public consciousness and to seek support for preventive medicine! What a basis to risk the high repute of the CDC! What a basis, for that matter, on which to expose 40 million people to an unknown risk of side effects! And all this on the word of experts, overconfident in theories validated through but two or three pandemics, without any proper review of their logic by disinterested scientists.
>
> It is not that the conclusions were inconsistent with the evidence, but that the *paucity of evidence* belied the force with which conclusions were advanced.[54]

That "basis" was a misuse of techniques as applied to a paucity of evidence.

Postscript

Policy Evaluation:
Art, Science, or Useless?

> *A free society requires free men and women who know what they are doing, who can make sense of their public lives by learning how to take effective action. Of what, then, does this rational action by citizens consist?*
>
> Aaron Wildavsky
> *Speaking Truth to Power: The Art and Craft of Policy Analysis*

Readers who have ventured this far may have lingering questions as to how policy is to be formulated, if not through the existing evaluative mechanism. If neither risk assessment nor cost-benefit analysis can be decisive, how *can* policy be implemented and assessed? If there is no scientific basis for our policies, what reasonable standards can be substituted for the judgment of scientific procedures?

These are important and controversial issues, and there can never be one simple, final set of answers. To begin, there is the political model itself; those who espouse this mechanism suggest that we ought to rely upon voters, acting through their representatives, to create and implement proper public policy. Such policy need never be set in stone as long as procedures exist through which our past policies can be overturned. When this occurs, it might be argued that such changes then reflect the will of the majority at the time; they are, therefore, defensible and desirable. This solution accepts the political process itself, not science, as the final arbiter—subject (in the United States) to judicial approval. It is in the courts, of course, that scientific rationales can enter even purely political disputes; courts are now free to decide cases on the basis of alleged scientific reasoning and experiments, regardless of whether purely scientific issues are at stake in such cases.[1]

The political model has a great deal of merit. The system of checks and balances, combined with adherence to the unencroachability of certain rights, has proven to be a valid and enduring model within which policy disputes can often be resolved. Its weaknesses are twofold: (1) the model relies on majoritarianism, and (2) the majority position, even so, can be overturned by judicial or political maneuvers (such as gerrymandering). Such outcomes are, of course, "good" or "bad" depending on one's own personal viewpoints. Evaluated as a policy system, however, it is difficult to deny the efficacy of this model. The search for any legitimate substitute is likely to be long and futile. All policy, finally, is normative; it will therefore be based on political considerations, regardless of any attempt to erect alternative, supposedly scientific models. Ethics, morality, political philosophy ...these are the proper foundations upon which to erect the edifice of public policies. Earl Warren's query about the difference between legal precedents and moral correctness (see note 1 above) goes to the heart of the matter; regardless of all the expert testimony, the *Brown v. Board of Education* decision was then (and remains now) the morally correct one— regardless of the ruling in *Plessy v. Ferguson,* even if that decision had stood for over half a century. There is no better way to decide such issues. Reliance on tradition, whether cultural or legal, is unworkable after cultural conditions have significantly changed. It is the virtue of a democratic political system that its mechanisms for change are available to those who wish to organize and use them.

What, however, of the tyranny of the majority? It was that possibility that led the framers of the American Constitution to remove from the political process certain fundamental prerequisites to the entire process itself; namely, those rights that could not be infringed by the federal government. This was as good a practical solution to the potential problem of a gradual loss of liberty as has been yet found on this planet. It also removed a set of conditions from the policy process taken to be axiomatically true and changeable only by that most conservative of majoritarian processes, a constitutional amendment. These declarations are purely normative in content. No cost-benefit study was done to examine whether or not the costs of allowing police to search at will outweighed the benefits of restricting them in that endeavor. No risk assessment study was completed to ascertain whether allowing people to keep firearms was excessively risky to themselves, their kin, or third parties. Such policies today would naturally require endless debate and input from social scientists; yet, it is doubtful that, whatever the outcome of that input, the resulting policies would be any better than those crafted solely from the normative belief structure of then-current political beliefs.

Because of the increasing complexity of our technology-driven society, it is almost always going to be the case that *someone* has to decide policy issues on some mutually-agreed-upon basis. Those decisions will require, and *ought to have,* the input of science and its practitioners. Even when the judiciary (or the set of juries within it) is the court of final appeal, expert testimony can never be the sole criterion for decision.[2] We typically place inordinate burdens on our judges and juries to decide highly complex issues requiring expertise beyond their existing abilities. Yet that is exactly the mechanism that was designed to deal with such issues. In the long run, the voters are ultimately empowered to alter the complexion of the nation's judiciary by altering the makeup of its legislatures and executives.[3] People will, regardless of the system's efficiency or ultimate justice, allow certain outcomes and forbid others. In so doing, they can be as wrong as any expert. Yet the "deliberate sense of the people" will remain the final judge of all policy issues.

The alternatives to such a model, although they might be more efficient and correct in the short run on particular policy proposals, are themselves fraught with potential dangers. Technocracy, central planning, and bureaucratic decree can produce tangible results, but they can never be successful substitutes for markets.[4] Further, those societies that are least willing to take risks are also those that are typically poor (and are going to remain so) relative to other societies willing to allow a higher degree of entrepreneurial freedom. We can easily manage risk by shutting down industrial activity altogether. The consequences of such policy, however, are, not pleasant to contemplate.

Our final appeal, then, is not to the judgments of risk authorities or those who claim to speak for the public interest, but to the public itself operating through its cherished political traditions. Societies that do not have such mechanisms for allocating risk and evaluating policy ought to erect them. When projected actions are seen as potentially conflicting with established legal canons, then only prudent judgments by Wildavsky's "rational citizens" can be decisive. Occasionally, those decisions will seem wrongheaded and needlessly anti-technology. But this is not a reflection of Luddism, but of the natural conservatism that arises from the broken promises and false certainties so often promulgated by expert policy analysts. There is usually nothing wrong with taking things more slowly in the face of known or unknown dangers. The overconfidence of experts can lead to horrifying consequences. The idea of unsinkable ships, be they passenger liners or scientific theories, is best abandoned in light of voluminous evidence to the contrary.

Normative policy judgments based on moral and political foundations will never be perfect or error-free. Neither, however, will be judgments by experts based on scientifically "positive" grounds. My grandmother used to say that "there was no damage like that often caused by educated fools." As I survey the entire terrain of our institutions with their expert-generated assumptions and often incredible outcomes, I begin to understand that my very uneducated relative was correct. There can never be any substitute for common sense and moral premises. Each time they are abandoned, humanity pays a terrible price. Science is a tool. It is not a method by which we can avoid making value judgments; it can never tell us what things are right and what things are wrong. Since public policy is about things happening to people, we must never lose sight of our normative roots, lest we end as a very sinkable ship in, ultimately, the very unnavigable waters of endless induction.

Prior normative criteria, however, are not always to be the final solution to policy questions. Widely accepted views, even when held for long periods of time, often turn out to be incorrect or in need of modification. Reasonable people will reflectively evaluate these criteria on a continuing basis and in light of new evidence that always becomes available over time. What cannot be done today may yet be done tomorrow for the simple reason that the future need not (and will not) be identical to the past. Change is inevitable. Our institutions allow for it to occur, precisely because of the certainty that it will occur. There is no reason to fear such change, as long as citizens remain rational and deliberate in their role as evaluators of the policies done in their name. People need not hold advanced degrees or be experts in natural or social science in order to assess accurately whether existing policy is good or bad, effective or counterproductive, useless or outdated. Their beliefs about the policies they live with are important, no matter how irrational they might appear to be to experts. Those beliefs will have to be dealt with in any event, so there is little reason for policy elites to ridicule them. Such attitudes are examples of the very unscientific mindset that experts so often deplore in other people, for they explicitly ignore reality by positing a world where perfectly-informed elites enact correct policies for the good of all, whether or not such policies are publicly supported. Such a world never has existed and never will—a fact that average people might respond to, however unscientifically, with a prayerful "Thank God."

Notes

Introduction

1. America is a nation derived from Enlightenment, i.e., rationalist traditions. As such, we have always placed great faith in science, especially as the fruits of technology became apparent. Pragmatically, it appeared that the triumphs in applied science could be easily transferred into areas where human actions were the issue. This has been attempted, with limited success, since the eighteenth century; it has now become a virtual obsession in the social sciences. It was begun by economists, but transcended that discipline, and is now the norm everywhere. Yet how well do the procedures of natural science transfer to the social sciences? Did the people who began this transfer actually understand what they were doing? These remain interesting and occasionally troubling issues in our time.

    See Mirowski (1984) for a discussion of the ignorance of nineteenth-century economists on the question of transferability of method.
2. Clark (1980) is excellent on the issue of finding what we are looking for. Also, see Rothwell (1982) on the same point for a discussion of definitions and their relationship to fact in the physical world.
3. Tracey (1978) quotes Alvin Weinberg on the definition of "transcience," the provision of answers to questions unanswerable by extant scientific techniques.
4. This contention may well surprise the reader. Any reasonably good statistics text should, no matter how elliptically, support the statement that statistics can prove nothing. Statistics, and statistical "tests," can be used to illustrate an argument, but never to prove one.
5. The silly contentions that infuse policy promises are well known to all Americans, from the electricity "too cheap to meter" to the end of poverty for the expending of "a mere $10 billion" to the "light at the end of the tunnel" that never came in Vietnam. Yet, along with the sillier hopes and political boilerplate, many more sophisticated, scientifically argued policies have fallen well short of expectations. Some plausible reasons for this are advanced in this book.
6. Bastiat (1968) suggested that the "state is that fiction by which each of us seeks to live at everyone else's expense." How does one improve upon that?
7. Some experts have always suffered from a lack of credibility. Economists, for example, have been both in and out of favor with the public. See Hutt's lament that their pronouncements are not taken as truth by laypersons in his day (1936). Today, even the statements of natural scientists are routinely disbelieved and even ridiculed. The reason suggested by this book is the general (and publicly perceived) corruption of science and its techniques by political special pleadings and desires for personal gain by practitioners. For a recent example, see S. Fred Singer's "My Adventures in the Ozone Layer," in *National Review*, June 30, 1989: 34-38.

8. Sowell (1980) brilliantly discusses our tendency to name our government agencies after "their hoped-for effects."
9. This is sometimes called the iron triangle theory of expenditure.

Chapter 1

1. Wildavsky and Douglas (1981) suggest that "whole bureaucratic and private financial empires" ride on such decisions.
2. Whelan (1985) suggests that the errors can be just as large when risk is overstated, which also undercuts public belief in the methods of analysts.
3. Lawless et al (1984).
4. Sowell (1980).
5. Lawless et al., p. 3.
6. Rowe (1977).
7. Kates (1978).
8. Lawless et al., p. 35.
9. Ibid., p. 27.
10. Ibid., p. 28.
11. Otway (1971); Beckman (1977); Joskow (1974).
12. Covello et al (1982).
13. Lawless et al., p. 24.
14. Ibid., p. 25.
15. Kates (1978).
16. Covello et al (1983).
17. Lawless et al., ch. 2.
18. Ibid.
19. Cummings (1981).
20. Schwing and Albers (1980).
21. Whittington and MacRae (1986).
22. Joskow (1974); Campen (1986).
23. This is sometimes a problem for new technologies where data is nonexistent, yet questions of risk are "answered" to comply with the law or soothe the public.
24. Clark (1980).
25. Semmons and Kresich (1989).
26. Ibid., p. 38.
27. Ibid., p. 39.
28. Efron (1984). The attempted poisoning of Rasputin is but one example of why toxicology is not quite an exact science.
29. Apostolakis (1978).
30. Excluding, of course, that the die does not behave in perverse ways (e.g., by landing on its edge)—that is, it is "fair."
31. While theories can indeed be "refuted" by evidence, no Classical probability predictions can ever be so refuted. This is because they are not theories but logically closed systems.
32. Popper (1972).
33. Just as no data can disprove or refute Classical probabilities, no data can prove them either. So all arguments that take the form "Las Vegas proves that Classical probability theory is true" are untenable.

34. Richard von Mises (1964), pp. 329-39. Bayes's Theorem can be stated as follows:

$$P(B/A) = \frac{P(A/B) \times P(B)}{P(A/B)P(B) + P(A/B_1)P(B_1)\ldots + P(A/B_n)P(B_n)}$$

Where $B_1 \ldots B_n$ are mutually exclusive and exhaust all possibilities. (Miller, p. 273)
35. Langlois (1980).
36. Hey (1983).
37. Richard von Mises (1964), p. 45. There is an old argument over whether or not probability theory is pure math or contains some inductive elements. The purely axiomatic constructions are associated with the names Kolmogorov, Cramer, Frechet, and Laplace. The theoretically verifiable *frequentist approach* is associated with the names Richard von Mises, Tornier, and Wald.
38. Kolata (1986).
39. Nalebuff (1988). The example is Jack Hirshleifer's.
40. Apostolakis (1978).
41. A perfect example of the difference between theoretical and actual would be to ask an expert contract bridge player what the "theoretically correct contract" for a particular combination of twenty-six cards (declarer and dummy) ought to be, based on *a priori* odds. That reply would not necessarily have much relevance to the empirically "correct" result of a particular hand actually dealt and played at the table. The reason is not simply the variations reality can assume; it is those variations in addition to the interactions of those distributions (facts) with human actions (subjective interpretations of possible facts). How this result can change even purely deductive analysis can be interesting. See, for example, *The Encyclopedia of Bridge,* 1984 ed., s.v. "Restricted Choice."
42. Perrow (1984).
43. Richard von Mises (1964), pp. 630-41.
44. Hey (1983).
45. My emphasis. Parry and Winter (1981), pp. 37-38.
46. Standing by the prediction is easier. No matter what happens in the physical world, no matter the "unlikeliness" of its occurrence, Classical methods allow for it and, in that sense, always "predict" reality as it is now.
47. Covello et al (1982).
48. Easterling (1981).
49. Covello et al (1983).
50. No outcome from the physical world can ever help us in determining whether or not a "die" is "fair"; therefore, "fairness" can never be inductively established. It remains a purely theoretic, *a priori,* category of thought.
    Actual frequency distributions, then, are a specific kind of knowledge, theoretically generated but containing factual content.
    See Shackle (1972), ch. 3.
51. Since Popperian falsificationism cannot be applied to the predictions of probability theory, then they are not "science" by that particular definition, which is the generally accepted standard for science today.
52. Inductive knowledge must have some use in this endeavor, since it is those outcomes we are attempting to model and/or predict. See Richard von Mises (1964), pp. 43-47.
53. Shackle (1972).

54. Ibid.
55. Ibid, ch. 32.
56. Hayek (1955, 1988).
57. Whelan (1985), ch. 3.
58. As cited by Whelan, ibid.
59. Cousins (1979).
60. Dickson (1984).
    The author was under contract to the National Science Foundation to design and evaluate a national survey of scientists. Two things of great and current concern to scientists generally were the "politicization of science" and "special pleading for political reasons" by fellow scientists. Further, scientists seemed generally to dislike the fact that their colleagues sometimes attempt to use the prestige science offers in their own field to comment, for perceived political reasons, in other fields and on the ongoing work in those fields by fellow scientists. Copies of the survey, entitled "A Survey of Scientists," are available from Sigma Xi, the National Honor Society of Scientists, New Haven, CT.
61. Cousins (1979).
62. Richard von Mises (1981).
63. WASH-1400/Nuclear Regulatory Commission's *Reactor Safety Study*. See also Beckman (1977).
64. McCraken (1982).
65. "Certain" by his own subjectively-held belief structure. This is, of course, quite controversial, even as it is widely practiced by both those calling themselves objectivist and subjectivist. See Easterling (1981) for the flavor of this debate.
66. One of the reported reasons for the delay in moving to clean up the oil spill from the Exxon Valdez tanker was a risk assessment that had predicted one such accident every 241 years. Because managers believed this assessment, they did not believe that this particular incident could be as bad as early reports had indicated.

Chapter 2

1. Wildavsky and Douglas (1981).
2. Menger (1976). English title: *Principles of Economics*.
3. White (1977), p. 3.
4. Ludwig von Mises (1984), p. 10.
5. Boehm-Bawerk (1959). Some of the bitterest criticisms were to come from the Austrians themselves, including Menger.
6. Wieser (1971).
7. Mises (1984), p. 17.
8. Ibid., p. 23. This argument was common. Hamilton wrote similarly concerning Adam Smith's ideas after the American Revolution. He praised Smith, but denied that his ideas on free trade applied to America. The Federalists were, after all, complete mercantilists.
9. Ibid.
10. Ibid, p. 25.
11. Ibid, p. 18.
12. Ibid, p. 28.
13. Ibid, p. 32.

14. Ibid.
15. Hayek (1955), pp. 198-99.
16. Mises was never to receive a full-time teaching job at any legitimate American university, an incredible loss both to Mises and the potential students he might have taught and influenced.

    Hayek was not in the economics department at Chicago because that department is one of America's militant Neoclassical strongholds. Hayek, although possessing impeccable credentials, just didn't "do" economics the way it was and still is done at Chicago.
17. Problems of both language and unfamiliar method were to plague Hayek in London. Joan Robinson recounts a question put to Hayek during one of his early LSE seminars on business cycles. The questioner asked whether, if he were to go out and purchase a raincoat, national income would actually *decline.* "Yes," replied Hayek, "but the mathematics is very difficult!" Indeed.
18. Some of the younger Austrians are Roger Garrison, Don Lavoie, Mario Rizzo, Gerald O'Driscoll, Richard Ebeling, Larry White, George Selgin, Jack High, Peter Lewin, Tyler Cowen, and Hans Hoppe.
19. Including the Public Choice school, located at George Mason University in Fairfax, Virginia and headed by Nobel laureate James Buchanan. One of Buchanan's early works, *Cost and Choice,* is explicitly Austrian. Another, edited with G.F. Thirlby and titled *LSE Essays on Cost,* contains many expositions by Austrians from across the Atlantic. McCloskey (1985) is a sympathetic Neoclassical.
20. White (1977), p. 1.
21. Hayek (1978), p. 3.
22. Blaug (1978), pp. 309-42; Ekelund and Hebert (1983), pp. 276-308; Schumpeter (1954), pp. 825-29; Simpson (1983), pp. 15-28; Mirowski (1984), pp. 361-79.
23. White (1977), p. 4.
24. Ibid., pp. 5-6.
25. Ibid., p. 4.
26. Menger (1976), pp. 188-92.
27. Rothbard (1956). Rothbard's contention is that there is, quite simply, *no such thing as total utility.* He writes:

    We must conclude that there is no such thing as total utility; all utilities are marginal. A typical error on the concept of marginal utility is a recent statement by Professor Kennedy that the word 'marginal' presupposes increments of utility and, hence, measurability. But the word 'marginal' presupposes *not* increments of utility, *but the utility of increments of goods,* and this need have nothing to do with measurability.
28. Stigler (1962), pp. 657, 665-66. Contrast with his treatment of Boehm-Bawerk (1968).
29. Stigler (1962), p. 670.
30. Ludwig von Mises (1966, 1978).
31. White (1977), p. 10.
32. Ludwig von Mises (1966), chs. 1 and 2.
33. White (1977), p. 9.
34. Ibid., p. 10.
35. Quoted by White, ibid.
36. Rothbard (1979).
37. Hayek (1955).

38. O'Driscoll (1977).
39. After Kuhn (1970). Because a "fact" requires prior definition, then all facts are colored by the theory we have of what they are and where we can find them.
40. Hayek (1948), pp. 33-76.
41. Hayek (1978), pp. 23-34.
42. White (1977), p. 15. Weber puts forth his methodology in the first part of his magnum opus (1978).
43. Quoted by White (1977), p. 16.
44. Schumpeter (1954); Stigler (1968).
45. Hutchison (1977). See also Lachmann (1976), pp. 65-80.
46. Lachmann (1976), pp. 152-59.
47. Macro variables are typically defined as being the average(s) of some set of data. This is necessary because some, such as price indices, simply do not exist and others, such as "the interest rate," must utilize a number that does duty for a whole panoply of various rates. These statistical constructs do not act on each other in the manner specified by macro models; hence they are of limited value in prediction or understanding.
48. Hollis and Nell (1975).
49. It was naively believed (as late as the 1950s) that empirical economic research would establish "facts" that all reasonable people could agree on, so that long-debated issues would at last be settled. That this has failed to happen—and the reasons that it has failed—might be gleaned from a reading of Kuhn (1970) and Feyerabend (1982).
50. Lavoie (1985). See also Morgenstern (1966).
51. Private conversation with Dr. Murray Rothbard, Las Vegas, NV. (October 3, 1986. Also present: Mrs. Joey Rothbard.) The flavor of the internecine argumentation can be gleaned from a reading of McCloskey's "Splenetic Rationalism," *Market Process,* vol. 7 (Spring, 1989), pp. 34-41.
52. Popper (1965, 1972).
53. "Better" because it has withstood many attempts to falsify it. See Friedman (1953) for support of this contention.
54. Popper (1972).
55. Hausman (1988).
56. Caldwell (1982), p. 13.
57. Rothbard (1979).
58. Admitted macroeconomic error possibilities can lead to rather amazing results, at least in theory. See Morgenstern (1966), pp. 242-76.
59. Hayek (1967a, 1967b).
60. If relationships can "shift," then prediction becomes implausible; some *ad hoc* model will always correlate with some past, lagged data set. This sort of science has been practiced by many economists other than Keynesians. See Miles (1984) for this criticism as it applies to Monetarism.
61. Hayek (1967a, 1967b).
62. Mackay (1932), ch. 1 is an entertaining chronicle of John Law's financial wizardry with the French macro-economy in the eighteenth century.
63. Blaug (1976).
64. See Adler (1985) and Kelley (1968) on this philosophical error.
65. Rothbard (1979).
66. My emphasis. Lavoie (1985), p. 57.
67. McCloskey (1985).

68. Lavoie (1985), p. 56.
69. Weaver (1948), p. 12.
70. Nutter (1983), p. 42.

Chapter 3

1. Campen (1986).
2. Hollis and Nell (1975).
3. Blaug (1987).
4. Hollis and Nell (1975).
5. Robbins (1935) is usually taken to be the source of this view.
6. Posner (1972).
7. New developments in quantum physics, chaos mathematics, and tendencies towards stable disequilibrium systems have implications, of course, for all static theorizing. Were the historicists partially right, but for the wrong reasons?
8. Browning and Browning (1988), ch. 8. Virtually any standard Neoclassical textbook would serve as well.
9. Kirzner (1973); Armentano (1972); Hayek (1948).
10. Market power, for Neoclassicals, is a firm's ability to charge a higher price without sacrificing all of its sales and therefore raise its revenues.
11. Browning and Browning (1988), pt. I. See also Mishan (1979), pp. 25-27.
12. Karl Menger (1973).
13. Armentano (1972), chs. 1 and 2.
14. Ibid.
15. Browning and Browning (1988), ch. 4; Friedman (1949); Bator (1954).
    Assumptions of infinite goods divisibility and zero wealth effects are sometimes left unstated, but apply nonetheless. Also, this applies only to Hicksian real income (or income compensated) demand functions.
    These modifications of Marshallian analysis lead to some interesting possibilities for CBA in the form of "duality theory," discussed below.
16. Mishan (1979); Campen (1986).
17. Mishan (1979). Shadow prices are implied by the theory of resource scarcity and, therefore, considerations of efficiency under competition. Yet it is precisely the concept of efficiency that is itself at issue. See note 65 below.
18. Samuelson (1974).
19. Ward (1979); Blaug (1978); Grossman (1962).
20. Sowell (1980).
21. Harberger (1969); Hirshleifer and Shapiro (1977); Baumol (1977); Arrow (1970).
22. Poole (1982).
23. Humbolt (1969).
24. "Facts" cannot tell us what to do in the absence of preexisting value-structures through which those facts are filtered and evaluated, including biases concerning what is possible at this time. See, for example, Philbrook (1953).
25. Wildavsky and Douglas (1982), pp. 30-31, on the fact-value issue.
26. Mansfield (1985), chs. 8, 14, and 15.
27. See, for example, Buchanan (1969).
28. Arrow (1966, 1977).
29. Harberger (1969), pp. 28-41.

30. Ibid., p. 40.
31. Musgrave and Musgrave (1989). See also Pasour (1988).
32. Arrow (1966); Baumol (1977).
33. Arrow (1966).
34. Governments redistribute rather than eliminate risk. Therefore, even govern-ments can become fiscally overburdened when events turn against them. Con-sider the savings and loan crisis in the United States.
35. My emphasis. Harberger (1969), p. 41.
36. Ibid.
37. Mises (1978); Rizzo (1978).
38. Static partial equilibrium analysis obscures the actual process of movements between positions. It occasionally traps itself in logical contradictions by seem-ingly forbidding such movements (all firms and individuals are price-takers) on the basis of its own assumptions. If all are price-takers, who can initiate the moves between positions?
39. Dynamic modeling has produced weak results, at best, in attempts to predict future economic states of the world. Donald McCloskey has pointed this out in an amusing fashion in his article "If You're So Smart, Why Aren't You Rich?" in *The American Scholar,* vol. 57 (Summer 1988), pp. 393-406.
40. They have corrected this tendency in recent years.
41. See Mises (1966) on the origin of interest and its nature as a fact of reality rather than an artifact of human design.
42. That is, the interbank loan rate. It is usually set below the prime rate on the grounds that banks are less risky borrowers. Today, the assumption still holds because of the existence of the Federal Reserve as a bail-out mechanism.
43. The point here is not whether governments build extremely risky things; cer-tainly, they often do. The point is: will they pay their liabilities? For many governments, that answer has been no.
44. This contention is no longer dismissed out of hand as it once would have been. No one really knows the total unfunded liabilities of the federal government, but estimates run as high as $12-15 trillion.
45. Miles (1984).
46. Mises (1966); Cowen and Fink (1985). The rate of interest that would prevail strictly due to the real factors in a market economy—pure time preference-based or not—would be, respectively, the originary rate of Mises and the natural rate of Wicksell.
47. Kalt (1981); Olsen (1965), pp. 9-15, 98-102.
48. Rothbard (1970), p. 835.
49. See, for example, the discussion of public goods in McConnell (1986).
50. See McCloskey (note 39, above). See also Morgenstern (1966); Rizzo (1978); and Lavoie (1985).
51. Browning and Browning (1979), pp. 205-6. They suggest that there is even an opportunity cost involved in the employment of unemployed resources.
52. Mansfield (1985), ch. 8. There is one problem with the Neoclassical analysis of the arrival mechanism for Pareto-optimality. When the initial endowment point is given within the Edgeworth Box, and the indifference curves passing through that point are not equal in slope (and hence in the marginal rates of substitution), the theory suggests that the two consumers will trade to a position consistent with Pareto-optimality.

This is fine as long as the movement is into the "lens-shaped" area created by the intersecting indifference curves. But texts also claim that a sliding move along each curve to a tangency position is an actual possibility to reach optimal distribution, since one consumer is better off and the other is no worse off.

This construction rests on the assumption that the consumer who is sliding along his curve is a pure altruist. This is never made clear in text discussions.

53. Lerner (1970) so argued, but see Blaug's criticism (1978), pp. 353-55.

54. Just et al. (1982), pp. 45-46.

55. This really is no different than the utilitarian approach taken when macro policy is considered. In theory, the aggregates are used for the sake of convenience only, but the costs of actually attempting to operationalize a summing of individual valuations would be prohibitive.

56. Bergson (1954).

57. Theorists who claim to know societal preferences can truly claim also to know the "public interest." But on what basis can such claims seriously be entertained?

58. Kuhn (1970) explains paradigms.

59. Lavoie (1987).

60. Mirowski (1984).

61. Mirowski (1988).

62. Friedman (1953).

63. Rothbard (1957).

64. Disputes over method typically are more bitter than those over ideology. A difference of agreement about "facts" at least holds out the possibility that one party will change his mind. What does one do, however, with an opponent who claims that your personal methods do not do what you claim and that you are wrong before you even reach a conclusion?

65. Heyne (1988). Austrians believe that technical and economic efficiency make no sense apart from values demonstrated in the marketplace.

66. Buchanan (1969), ch. 3.

67. Ibid., chs. 1 and 2.

68. Is subjectivism smuggled into the Neoclassical analytic? It is in a strange way. See Alchian (1977) for the foundations of that insertion.

69. Armentano (1988).

70. Blaug (1987), pp. 159-62.

71. Rothbard (1956) quotes Ellsberg's attack on the concept of measurable utility (von Neumann and Morgenstern's famous 1947 work) approvingly. In fact, he is correct that ordinality and subjectivity have carried the day in economics. But see Cooter and Rappaport (1984) and Blaug's examination of introspection (1987).

72. Money prices will not suffice because of the excess of utility over money actually sacrificed at the time of the trade.

73. Rothbard (1956).

74. Rothbard, ibid; Samuelson (1982), chs. 1, 9, and 10. At first glance, revealed and demonstrated preference appear identical. Yet revealed preference requires a constancy of underlying value scales that demonstrated preference does not. The attempt to operationalize revealed preferences in CBA is addressed in note 76 below. Both revealed and demonstrated preferences are attempts to avoid the earlier questionnaire approach as well as the pure ordinalist, constant-value analysis in Hicks (1946).

75. Hicks (1946), chs. 1 and 2.
76. Baumol (1977b); (1984), and Just et al. (1982) have analyses of "duality theory." Instead of viewing a consumer as maximizing utility within a set of expenditure (income) constraints, duality theory approaches the question from the expenditure side itself. Consumers are seen as minimizing expenditures for the attainment of some given level of utility. In this way, pure income-compensated demand curves are estimated, thus bypassing all income effects when prices and quantities demanded are matched and thus providing theoretically an almost perfect measure of benefits.

   All this is nice in theory, but the requisite information cannot be generated without operationally large expenditures. Further, this approach has to come back to verbal-based estimations for the simple reason that preferences change over time, so that observations—even if they could be performed in sufficient numbers—would not necessarily be uniform when translated into the demand functions.
77. Samuelson (1974), ch. 6.
78. Rothbard (1956), pp. 26-27.
79. Quoted by Rothbard, ibid., p. 10.
80. Mises (1978); Rizzo (1978).
81. Hoehn et al. (1989) suggest that the passage of time and the distributional effects of redistributive policies make the benefits being added as incomparable as apples and oranges.
82. Kirzner (1973); Armentano (1972), chs. 1-2.
83. Its seeming lack of concern for the replacement of infrastructure until it crumbles—as is the case in our major cities, with our roads, and in governmentally-run weapons plants—is but one of dozens of examples that could be cited in support of this contention.
84. Rothbard (1979).

Chapter 4

1. All policy creates winners and losers. The argument that "society" has gained when the net gains of winners are greater than the net losses of losers is simply a restatement of utilitarian dogma ("greatest good for the greatest number") along with an untenable reification of the entity "society."
2. Campen (1986) discusses whether politics interferes with CBA. See also Schwartz (1986), pp. 82-85, as well as Dickson (1984).
3. It certainly did fail according to the public's understanding, and perhaps that is the final arbiter of such matters.
4. Popper (1972); Blaug (1987), pp. 29-54.
5. "Normative" implies values, while "positive" implies only facts. The distinction, common in modern "scientific policy analysis," comes from the positivist episode in philosophy.
6. As quoted in Burnham (1965), p. 54.
7. Although Wildavsky and Douglas (1981) suggest that it is the public who argue these matters correctly, while it is the analysts who commit "ancient logical fallacies."
8. Mises (1966) and Simon (1976).
9. Neustadt and Fineberg (1978).

10. The "right thing," within this context and given the methodology of scientists, must refer to either 1) the correct political moves within the context of swine flu decision-making, or 2) "things that can be seen as right" from some theological viewpoint.
11. This chronology is principally a paraphrase of the one supplied by Neustadt and Fineberg (1978). Also, there is some information from Silverstein (1981).
12. These conditions are known personally by the author, who completed basic training there between January and March of 1969. Hundreds of flu cases at Fort Dix during this time of year under those sorts of conditions were typical. Treatment was nonexistent, and forced marches of infected recruits was also a common occurrence.
13. Silverstein (1981).
14. A confusion between "verbal-based preferences" and "action-based preferences"? (See Chapter Three.)
15. Given that no program of this magnitude had, as yet, been undertaken, this statement must have been pure propaganda. How, then, did the CDC calculate this "range"? This is one of the more disturbing aspects of government medicine—the tendency to lie when the truth would be useful to the public but detrimental to certain bureaucratic careers.
16. This silliness is perfectly analogous to President Carter's visit to Three Mile Island during its famous nuclear accident. In each case, the official position was that there was no danger to the chief executive, and millions watched and were reassured. Yet, in each case, the official position was wrong, and those advising the president surely suspected it might be wrong.
17. "Serious criticism" is that taken seriously enough by the media to warrant large dissemination. Although there are people (the author is one) who would have criticized this program on the fundamental grounds of state intervention, it is doubtful that such a position would have been taken seriously either then or now.
18. Schoenbaum et al (1976).
19. Ludwig von Mises (1963).
20. Neustadt and Fineberg (1978), p. 10.
21. Ibid., p. 11.
22. Murray (1988). There is absolutely nothing "unscientific" in this method; in Murray's case, his use of it leads the reader into some insightful lines of thought and argumentation.
23. One participant told Neustadt and Fineberg (1978), p. 25:

    There was no way to go back after Sencer's memo (to Ford). If we tried to do that, it would leak. *That memo was a gun to our heads.* [My emphasis.]
24. Rowley et al (1988).
25. Kilbourne (1975).
26. Silverstein (1981). Ironically, older people would typically have been placed (and it was argued that they *ought* to be placed) within a so-called "high risk" group. Yet they would have been the group that had some natural immunity to the disease.
27. My emphasis. Yes, it was greater than zero; so what? This fact leads to no one conclusion or program alternative, or even to the alleged necessity of acting at all.
28. Neustadt and Fineberg (1978), p. 11.
29. My emphasis. Silverstein (1981), p. 41.

30. Neustadt and Fineberg (1978), p. 16.
31. Arrow and Lind (1970) provide a useful overview of this argument and some criticisms.
32. Ludwig von Mises (1963).
33. Neustadt and Fineberg (1978), p. 10.
34. The theory is Kilbourne's (1975). If the theory had been true, then it was already too late for the CDC's ambitious program. But if not, then there was no reason not to stockpile as Sabin had suggested.
35. Silverstein (1981), p. 4.
36. Kilbourne (1975).
37. Byrns and Stone (1986), p. 153.
38. As related in Schoenbaum et al (1976). Also, Neustadt and Fineberg (1978), p. 25.
39. Silverstein (1981), p. 62.
40. Behn and Vaupel (1982).
41. Schoenbaum et al (1976).
42. Ibid.
43. Milholland et al (1978).
44. Schoenbaum (1976), p. 760.
45. Ibid.
46. Ibid., p. 761.
47. Ibid.
48. These cannot be legitimately used as a basis for arguing for state intervention in order to "correct for them." At least, most economists do not so argue.
49. This is but another expression of the traps we set for ourselves when we argue about collectivities and when we aggregate "data." Clearly, arguments become cosmic and therefore untenable when such godlike judgments are required.
50. *Historical Statistics of the United States: Colonial Times to 1970* (Washington: U.S. Department of Commerce, Bureau of the Census, 1972).
51. The implicit assumption that the natures of benefits and costs are homogeneous is a vast simplification of reality. In fact, the aggregation of "costs" (which are, after all, merely choices made at various moments in time) is misleading and probably impossible. The data cannot be collected, as it is the subjectively forgone opportunities of individuals. Where do data sets like these exist?
52. The "answers" given by current tort cases vary as the system attempts to sift more information about the way that people value things. Nonetheless, any technique will always fail to satisfy some groups of individuals as to its "correctness." See Barrett (1988).
53. As quoted by Neustadt and Fineberg (1978), p. 71.
54. My emphasis. Ibid, p. 101.

Postscript

1. Sowell (1980), ch. 9. The scientific evidence presented in *Brown v. Board of Education* was merely window dressing for the final outcome. Yet Earl Warren's comment from the bench—"Yes, that's the law, but is it *right*?"—serves as the focal point for understanding the decision.

   There are, of course, large costs involved every time a precedent is overturned. In this case, the decision led to busing and other policies that violated the original intent of the decision. (Yes, even isolated legal decisions obviously have original intent.) But that is beside the point. Enduring error needs correction, and it will be corrected—if not peaceably, then otherwise.

2. See Wayne E. Green, "Expert-Witness Decisions Hurt Consumer Toxic Cases," in *The Wall Street Journal,* p. B1, July 11, 1989.

3. In states such as California, judges can also be changed through recall elections. Although this seldom happens, when the people do rise and vote the result can be dramatic.

4. Hayek (1975) and Lavoie (1985a) provide a good overview of the issues. What is "perestroika" if not a tacit admission of their arguments?

Bibliography

Adler, Mortimer J. (1985) *Ten Philosophical Mistakes* (New York: MacMillan).

Alchian, Armen (1977) "Uncertainty, Evolution, and Economic Theory," in *Economic Forces At Work* (Indianapolis: Liberty Press), 15-36.

Ames, Bruce N., Renae Magaw, and Lois S. Gold (1987) "Ranking Possible Carcinogenic Hazards," *Science,* vol. 236 (April 17, 1987), pp. 261-80.

Apostolakis, George (1978) "Probability and Risk Assessment: The Subjectivist Viewpoint and Suggestions," *Nuclear Safety,* vol. 19, no. 3 (May-June, 1978).

Armentano, D. T. (1972) *The Myths of Antitrust* (New Rochelle: Arlington House).

---------- (1988) "Rothbardian Monopoly Theory and Antitrust Policy," in Block and Rockwell (eds.), pp. 3-11.

Arrow, Kenneth (1966) "Discounting and Public Investment Criteria," in Water Resources Research (A. V. Kneese and S. Smith, Johns Hopkins University Press).

---------- (1977) *Studies in Resource Allocation Processes* (Cambridge: Cambridge University Press).

---------- (1983) *Social Choice and Justice* (Cambridge: Belknap Press-Havard University Press).

Arrow, Kenneth, and R. C. Lind (1970) "Uncertainty and the Evaluation of Public Investment Decisions," *American Economic Review,* vol. LX, no. 3 (June 1970), pp. 364-78.

Ayer, A. J. (1972) *Probability and Evidence* (New York: Columbia University Press).

Bailey, Martin J. (1954) "The Marshallian Demand Curve," *Journal of Political Economy* (June, 1954). Reprinted in Breit and Hochman (eds.), pp. 103-110.

Bailey, Martin J., and Michael Jensen (1972) "Risk and the Discount Rate for Public Investment." *Studies in the Theory of Capital Markets* (New York: Praeger).

Bails, Dale G., and Larry Peppers (1982) *Business Fluctuations: Forecasting Techniques and Applications* (Englewood Cliffs: Pretice-Hall).

Barrett, Paul (1988) "New Legal Theorists Attach A Value to the Joys of Living," *Wall Street Journal* (December 12, 1988), p. A1.

Bastiat, Fredric (1968) *The Law* (Irvington-on-Hudson: Foundation for Economic Education).

Bator, Francis (1957) "The Simple Analytics of Welfare Maximization," in Breit and Hochman (eds.), pp. 455-83.

113

---------- (1988) "The Anatomy of Market Failure," in Tyler Cowen (ed.), p. 35-68.

Baumol, William J. (1977a) "On The Discount Rate for Public Projects," in Haveman and Margolis (eds.), pp. 161-79.

---------- (1977b) *Economic Theory and Operations Analysis* (Englewood Cliffs: Prentice-Hall).

---------- (1984) "Baumol on Hicks," in Henry Spiegel and Warren Samuels (eds.), pp. 37-64.

---------- (1986) *Superfairness: Applications and Theory* (Cambridge: MIT Press).

Beckman, Peter (1977) *The Health Hazards of Not Going Nuclear* (Boulder: Golem Press).

Behn, Robert D., and James W. Vaupel (1982) *Quick Analysis for Busy Decision Makers* (New York: Basic Books).

Berger, James O., and Donald Berry (1988) "Statistical Analysis and the Illusion of Objectivity," *American Scientist,* vol. 76 (March-April, 1988), pp. 159-65.

Bergson, Abram (1954) "On the Concept of Social Welfare," *Quarterly Journal of Economics* (May, 1954), pp. 249-61.

---------- (1975) "A Note on Consumer's Surplus," *Journal of Economic Literature,* vol. 13, no. 1 (March, 1975), pp. 38-44.

Blalock, Herbert (1979) *Social Statistics* (New York: McGraw-Hill).

Blanshard, Brand (1969) *The Nature of Thought* (London: George Allen and Unwin Ltd).

Blaug, Mark (1976) *The Cambridge Revolution* (London: Institute for Economic Analysis).

---------- (1978; 3rd ed.) *Economic Theory in Retrospect* (Cambridge: Cambridge University Press).

---------- (1986) *Economic History and the History of Economics* (New York: New York University Press).

---------- (1987) *The Methodology of Economics: How Economists Explain* (Cambridge: Cambridge University Press).

Block, Walter, and Llewellyn Rockwell (eds.) (1988) *Man, Economy, and Liberty: Essays in Honor of Murray Rothbard* (Auburn, Ludwig von Mises Institute).

Boetke, Peter, and Steven Horowitz (1986) "Beyond Equilibrium Economics: Reflections on the Uniqueness of the Austrian Tradition," *Market Process,* vol. 4, no. 2 (Fall 1986).

Borch, Karl Hendrick (1986) *The Economics of Uncertainty* (Princeton: Princeton University Press).

Bowen, Earl (1967) *Mathematics With Applications in Management and Economics* (Homewood: Richard Irwin).

Boehm-Bawerk, Eugen von (1959) *Capital and Interest* (South Holland: Libertarian Press).

Breit, William, and Harold Hochman (eds.) (1971) *Readings in Microeconomics* (Hinsdale: Dryden Press).

Browning, Edgar (1988; 3rd ed.) *Microeconomics* (Englewood Cliffs: Prentice Hall).

Browning, Edgar and Jacqueline Browning (1979) *Public Finance and the Price System* (New York: MacMillan).

Braybrooke, D., and C. E. Lindbloom (1963) *A Strategy of Decision: Policy Evaluation As A Social Process* (New York: The Free Press).

Buchanan, James (1969) *Cost and Choice* (Chicago: Markham Publishers).

---------- (1975) *The Limits of Liberty* (Chicago: University of Chicago Press).

---------- (1982) "The Domain of Subjective Economics: Between Predictive Science and Moral Philosophy," in Israel Kirzner (ed.) pp. 7-20.

Buchanan, James, and G. F. Thirlby (eds.) (1973) *LSE Essays on Cost* (London: Weidenfield and Nicholson).

Buchanan, James, and Gordon Tullock (1974) *Calculus of Consent: The Logical Foundations of Constitutional Democracy* (Ann Arbor: University of Michigan Press).

Burnham, James (1965) *Suicide of the West* (New Rochelle: Arlington House).

Byrns, Ralph T., and Gerald W. Stone (1986) *Economics* (New York: Scott Foresman and Company).

Cadwell, Bruce (1982) *Beyond Positivism* (Boston: Allen and Unwin).

---------- (1988) "Hayek's Transformation," *History of Political Economy,* vol. 20, no. 4 (Winter, 1988), pp. 513-41.

Campen, James T. (1986) *Benefit, Cost and Beyond: The Political Economy of Benefit-Cost Analysis* (Cambridge: Ballenger).

Carnap, Ruldolph (1962) *Logical Foundations of Probability* (Chicago: University of Chicago Press).

Clark, William C. (1980) "Witches, Floods, and Wonder Drugs," in Schwing and Albers (eds.), pp. 284-311.

Clower, Robert (1984) "Monetary History and Positive Economics," in *Money and Markets: Essays by Robert Clower.* Donald A. Walker (ed.) (Cambridge: Cambridge University Press).

Coffey, P. (1958) *Epistemology: An Introduction to General Metaphysics* (Gloucester, MA: Peter Smith).

Cooter, Robert and Peter Rappaport (1984) "Were the Ordinalists Wrong?" *Journal of Economic Literature,* vol. 22, no. 2 (June, 1984), pp. 507-30.

Cousins, Norman (1979) *Anatomy of An Illness As Perceived By the Patient* (New York: Norton).

Covello, V.T., J. Menkes, and J. Nehnevajsa (1982) "Risk Analysis, Philosophy, and the Social and Behavioral Sciences: Reflections on the Scope of Risk Analysis Research," *Risk Analysis,* vol. 2, no. 2 (June, 1984).

Covello, V. T., W. G. Flamm, et al. (1983) *The Analysis of Actual Versus Perceived Risk* (New York: Plenum Press).

Cowen, Tyler (ed.) (1988) *The Theory of Market Failure: A Critical Examination* (Fairfax, VA: George Mason University Press/Cato Institute).

Cowen, Tyler and Richard Fink (1985) "Inconsistent Equilibrium Constructs: The ERE of Mises and Rothbard," *American Economic Review*, vol. 75, no. 4 (September, 1985).

Cummings, R. B. (1981) "Is Risk Assessment A Science?", *Risk Analysis*, vol. 1, no. 1 (March, 1981).

Dahlman, Carl J. (1988) "The Problem of Externality," in Tyler Cowen (ed.), pp. 209-36.

Datson, Lorraine (1987) "The Domestication of Risk: Mathematical Probability and Insurance," in Datson et al. (eds.) vol. 1, pp. 237-60.

Datson, Lorraine, Lorenz Kruger, and Michael Heidelberger (eds.) (1987) *The Probabilistic Revolution* (Cambridge, MA: MIT Press).

Denton, Michael (1986) *Evolution: A Theory in Crisis* (Bethesda, MA: Adler and Adler).

Dickson, David (1984) *The New Politics of Science* (New York: Pantheon).

Douglas, R. Gordon (1975) "Influenza in Man," in Edwin Kilbourne (ed.), pp. 412-23.

Dreze, J. H. (1974) *Allocation Under Uncertainty: Equilibrium and Optimality* (London: MacMillan).

Easterday, B. C. (1975) "Animal Influenza," in Edwin Kilbourne (ed.), pp. 464-68.

Easterling, Robert (1981) "Special Letter," in *Nuclear Safety*, vol. 22, no. 4 (July-August, 1981), pp. 464-65.

Efron, Edith (1984) *The Apocalyptics* (New York: Simon and Schuster).

Ekelund, Robert and Robert Hebert (1983) *A History of Economic Theory and Method* (2nd ed.) (New York: McGraw-Hill).

Ekirch, Arthur A. (1974) *Progressivism in America* (New York: New Viewpoints).

Feyerabend, Paul (1982) *Against Method* (London: Redwood Burn).

Fiske, Donald W. and Richard A. Shweder (eds.) (1986) *Metatheory in Social Science: Pluralisms and Subjectivities* (Berkeley: University of California Press).

Freudenburg, William R. (1988) "Perceived Risk and Real Risk: Social Science and the Art of Probabilistic Risk Assessment," in *Science*, vol. 242, no. 4875 (October, 1988) pp. 44-49.

Friedman, Milton (1949) "The Marshallian Demand Curve," *Journal of Political Economy* (December). Reprinted in Breit and Hochman, 1971, pp. 92-102.

---------- (1953) "The Methodology of Positive Economics," in *Essays in Positive Economics* (Chicago: University of Chicago Press), pp. 3-46.

---------- (1962a) "Is A Free Society Stable?" in *New Individualist Review* vol. 2, no. 2 (Summer, 1962), pp. 3-10.

---------- (1962b) "Leon Walras and His Economic System," in Spengler and Allen (eds.), pp. 672-81.

Friedrich, Carl J. (ed.) (1977) *The Philosophy of Kant* (New York: The Modern Library).

Froyen, R. T. (1985) *Macroeconomics: Theories and Policies* (New York: McGraw-Hill).

Gillies, D. A. (1973) *An Objective Theory of Probability* (London: Methuen).

Glymour, Clark (1980) *Theory and Evidence* (Princeton: Princeton University Press).

Goodman, John C. (1986) "Privatization in Great Britain," in *Wall Street Journal,* October 12, editorial page.

Gough, Michael (1986) *Dioxin, Agent Orange: The Facts* (New York: Plenum Press).

Grassl, Wolfgang and Barry Smith (eds.) (1986) *Austrian Economics* (New York: New York University Press).

Grossman, Henryk (1962) "The Evolutionist Revolt Against Classical Economics," in Spengler and Allen (eds.), pp. 500-24.

Gwartney, James and Richard Wagner (eds.) (1988) *Public Choice and Constitutional Economics* (Washington: Cato Institute).

Hahn, Frank and Martin Hollis (1979) *Philosophy and Economic Theory* (Oxford: Oxford University Press).

Harberger, Arnold C. (1969) *Project Evaluation* (Chicago: University of Chicago Press).

Hausman, Daniel (1988) "An Appraisal of Popperian Methodology," in *The Popperian Legacy in Economics,* Neil DiMarchi (ed.) (Cambridge: Cambridge University Press), pp. 65-86.

---------- (1989) "Economic Methodology in A Nutshell," *Journal of Economic Perspectives,* vol. 3, no. 2 (Spring, 1989), pp. 115-28.

Hayek, F. A. (1948) "Facts of the Social Sciences" and "Economics and Knowledge," in *Individualism and Economic Order* (Chicago: University of Chicago Press), pp. 33-76.

---------- (1955) *The Counterrevolution of Science* (London: Collier-MacMillan/ Glencoe Paperback).

---------- (1967a) *Monetary Theory and the Trade Cycle* (New York: Augustus Kelley).

---------- (1967b) *Prices and Production* (New York: Augustus Kelley).

---------- (1975) *Collectivist Economic Planning* (Clifton, NJ: Augustus Kelley).

---------- (1978) "The Pretense of Knowledge," in *New Studies in Politics, Economics and Philosophy* (Chicago: University of Chicago Press), pp. 23-34.

---------- (1984) "Marginal Utility and Economic Calculation," in *Money, Capital, and Fluctuations* (Chicago: University of Chicago Press), pp. 183-89.

---------- (1988) *The Fatal Conceit* (Chicago, IL: University of Chicago Press).

Herbert, Nick (1985) *Quantum Reality* (New York: Anchor Books/Doubleday).

Hey, John D. (1979) *Uncertainty in Microeconomics* (New York: New York University Press).

---------- (1983) *Data in Doubt: An Introduction to Bayesian Statistical Inference for Economists* (Oxford: Martin Robinson Press).

Hicks, Sir John (1946) *Value and Capital* (2nd ed.) (London: Oxford University Press).

---------- (1979) *Causality in Economics* (New York: Basic Books).

Hirshleifer, Jack and John Riley (1979) "The Analytics of Uncertainty and Information: An Expository Survey," *Journal of Economic Literature,* vol. 17, no. 4 (December, 1979), pp. 1375-1421.

Hirshleifer, Jack and David L. Shapiro (1977) "The Treatment of Risk and Uncertainty," in Haveman and Margolis (eds.), pp. 180-203.

Hoehn, John P. and A. Randall (1989) "Too Many Proposals Pass the Benefit-Cost Test," *American Economic Review,* vol. 79, no. 3 (June, 1989), pp. 544-51.

Hoff, T. J. B. (1949) *Economic Calculation in the Socialist Society* (London: William Hodge and Company).

Hollis, Martin and Edward Nell (1975) *Rational Economic Man* (Cambridge: Cambridge University Press).

Hoppe, Hans-Hermann (1988) *Praxeology and Economic Science* (Auburn, AL: Ludwig von Mises Institute).

Horwich, Peter (1982) *Probability and Evidence* (Cambridge: Cambridge University Press).

Humbolt, Wilhelm von (1969) *The Limits of State Action* (Cambridge: Cambridge University Press).

Hutchison, T. W. (1977) *Knowledge and Ignorance in Economics* (Chicago: University of Chicago Press).

---------- (1981) *The Politics and Philosophy of Economics* (New York: New York University Press).

Hutt, William H. (1936) *Economists and the Public* (London: Jonathan Cape).

Jacobs, Jane (1984) *Cities and the Wealth of Nations* (New York: Random House).

Jeffrey, R. C. (1983) "Probability and Falsification: A Critique of the Popperian Programme," in *Methodologies: Bayesian and Popperian,* Synthese, vol. 30.

Joskow, P. L. (1974) "Approving Nuclear Power Plants: Scientific Decisionmaking or Administrative Charade?" *Bell Journal,* vol. 5, no. 2 (Spring, 1974).

Just, Richard, Darrell Hueth, and Andrew Schmitz (1982) *Applied Welfare Economics and Public Policy* (Englewood Cliffs, NJ: Prentice-Hall).

Kalt, Joseph P. (1981) "Public Goods and the Theory of Government," *Cato Journal,* vol. 1, no. 1 (Fall, 1981), pp. 565-84.

Kamlah, Andreas (1987) "The Decline of the Laplacian Theory of Probability," in Datson et al (eds.), pp. 91-116.

Kates, R. W. (1978) *Risk Assessment and Environmental Hazard* (New York: John Wiley).

Kelley, David (1988) *The Evidence of the Senses* (Baton Rouge, LA: Louisiana University Press).

Keynes, John Maynard (1973) *A Treatise on Probability* (New York: St. Martin's Press). (Vol. 8 of the Royal Academy of Science's Collected Works).

Kilbourne, Edwin D. (1975) "Epidemiology and Influenza," in *The Influenza Viruses and Influenza,* Edwin Kilbourne (ed.), pp. 483-538.

---------- (ed.) (1975) *The Influenza Viruses and Influenza* (New York: Academic Press).

Kirzner, Israel (1973) *Competition and Entrepreneurship* (Chicago: University of Chicago Press).

---------- (1978) "Economics and Error," in *New Directions in Austrian Economics,* Louis Spadaro (ed.) (Kansas City: Sheed, Andrews, and McNeel).

---------- (ed.) (1982) *Method, Process, and Austrian Economics: Essays in Honor of Ludwig von Mises.* (Lexington, MA: D.C. Heath and Company).

Klamer, Arjo, Don McCloskey and R. Solow (eds.) (1988) *The Consequences of Rhetoric* (Cambridge University Press: Cambridge University Press).

Knight, Frank (1971) *Risk, Uncertainty, and Profit* (Chicago: University of Chicago Press).

Kolata, Gina (1986) "What Does It Mean to Be Random?" *Science,* vol. 231, March 7, pp. 1068-70.

Kuhn, Thomas (1970) *The Structure of Scientific Revolutions* (Chicago: University of Chicago Press).

---------- (1987) "What Are Scientific Revolutions?" in Datson et al (eds.), pp. 7-22.

Lachmann, Ludwig (1976a) "Towards A Critique of Macroeconomics," in *The Foundations of Modern Austrian Economics* (Kansas City: Sheed and Ward).

---------- (1976b) "From Mises to Shackle: An Essay," *Journal of Economic Literature,* vol. 14, no. 1 (March, 1976), pp. 54-62.

---------- (1977) *Capital, Expectations, and the Market Process* (Kansas City: Sheed, Andrews, and McNeel).

---------- (1982) "Ludwig von Mises and the Extension of Subjectivism," in Israel Kirzner (ed.), pp. 31-40.

Lagerspetz, Eerik (1984) "Money as A Social Contract," *Theory and Decision,* vol. 17, no. 1 (July, 1984), pp. 1-9.

Lakatos, Imre (1970) "Falsification and the Methodology of Scientific Research Programmes," in *Criticism and the Growth of Knowledge,* Imre Lakatos and Alan Musgrave (eds.) (Cambridge: Cambridge University Press).

Lange, Oscar (1964) *On the Economic Theory of Socialism* (New York: McGraw-Hill).

Langlois, Richard (1980) "Subjective Probability in Austrian Economics," unpublished paper.

---------- (1982a) "Cost-Benefit Analysis, Environmentalism and Rights," *Cato Journal,* vol. 2, no. 1, pp. 279-301.

---------- (1982b) "Austrian Economics As Affirmative Science," in Kirzner (ed.), pp. 75-84.

Lavoie, Don (1985a) *Economic Planning: What Is Left?* (Washington: Cato Institute).

---------- (1985b) "The Interpretive Dimension in Economic Science: Hermeneutics and Praxeology" (Center for Study of Market Process, George Mason University) Working Paper Series, Number 15.

---------- (1985) "From Hollis and Nell to Mises," *Journal of Libertarian Studies,* vol. 1, no. 4, pp. 325-36.

Lawless, Edward, Martin Jones, and Richard Jones (1985) *Comparative Risk Assessment* (Kansas City: Midwest Research Institute). (A National Science Foundation Final Report).

Lerner, Abba P. (1970) *The Economics of Control* (New York: Augustus Kelley).

Lindbloom, C. E. (1986) *The Policymaking Process* (Englewood Cliffs, NJ: Prentice Hall).

Lipkin, Richard (1988) "Risky Business of Assessing Danger," *Insight,* May 23, pp. 8-16.

Littlechild, Stephen C. (1979) "The Problem of Social Cost," in Spadaro (ed.), pp. 77-93.

Loeb, Louis E. (1981) *From Descartes to Hume: Continental Metaphysics and the Development of Modern Philosophy* (Ithaca, NY: Cornell University Press).

Lowrance, William O. (1976) *Of Acceptable Risk: Science and the Determination of Safety* (Los Altos, CA: William Kaufman).

Machina, Mark J. (1987) "Decision-Making in the Presence of Risk," *Science,* May 1, pp. 537-42.

Machlup, Fritz (1976) *Selected Writings of Fritz Machlup,* George Britos (ed.). (New York: New York University Press).

---------- (1978) "Theories of the Firm: Marginal, Behaviorial, and Managerial," in *Methodology of Economics and Other Social Sciences* (New York: Academic Press).

---------- (1984) *Knowledge: Its Creation, Distribution, and Economic Significance* (Princeton: Princeton University Press), vol. 3.

Mackay, Charles (1841) *Extraordinary Popular Delusions and the Madness of Crowds* (London: Richard Bentley). (Reprint of original Farra, Straw, and Giroux, 1932).

Mansfield, Edwin (1985) *Microeconomics* (New York: Norton).

Marshall, Eliot (1985) "The Academy Kills A Nutrition Report," *Science,* vol. 230, (October 25, 1985), pp. 230-31.

McCloskey, Donald (1985) *The Rhetoric of Economics* (Madison: University of Wisconsin Press).

McCormack, Norman J. (1981) *Probability, Risk Analysis Methods and Nuclear Power* (New York: Academic Press).

McCraken, Samuel (1982) *The War Against the Atom* (New York: Basic Books).

McNeil, William (1976) *Plagues and Peoples* (New York: Doubleday).

Menard, Claude (1987) "Why Was There No Probabilistic Revolution in Economics?" in Datson et al (eds.), pp. 139-46.

Menger, Carl (1976) *Principles of Economics* (New York: New York University Press).

---------- (1985) *Investigations Into the Method of the Social Sciences With Special Reference to Economics,* Louis Schneider (ed.) (New York: New York University Press).

Menger, Karl (1973) "Austrian Marginalism and Mathematical Economics," in *Carl Menger and the Austrian School* (Oxford: Clarendon Press).

Miles, Marc (1984). *Beyond Monetarism* (New York: Basic Books).

Milholland, Arthur, Stanley Wheeler, and J. Heieck (1973) "Medical Assessment by A Delphi Group Opinion Technique," *New England Journal of Medicine,* vol. 288, no. 24 (June 14, 1973), pp. 1272-1274.

Miller, Richard (1987) *Fact and Method: Explanation, Confirmation, and Reality in the Natural and Social Sciences* (Princeton: Princeton University Press).

Mirowski, Philip (1984) "Physics and the Marginalist Revolution," *Cambridge Journal of Economics,* vol. 8, no. 4.

---------- (1988) "Rhetoric, Mathematics, and the Nature of Neoclassical Theory," in Klamer et al. (ed.), pp. 117-45.

Mises, Ludwig von (1963) *Bureaucracy* (New Rochelle, NY: Arlington House).

---------- (1966) *Human Action* (Chicago: Regnery).

---------- (1978) *The Ultimate Foundation of Economic Science* (Kansas City: Sheed, Andrews and McNeell).

---------- (1984) *The Historical Setting of the Austrian School* (Auburn, AL: The Ludwig von Mises Institute).

Mises, Richard von (1965) *The Mathematical Theory of Probability and Statistics* (New York: Academic Press).

---------- (1981) *Probability, Statistics and Truth* (New York: Dover Paperbacks).

Mishan, E. J. (1952) "The Principle of Compensation Reconsidered," *Journal of Political Economy,* vol. 60 (August, 1952), pp. 312-322.

---------- (1979) *Cost Benefit Analysis* (London: George Allen and Unwin).

Morgenstern, Oscar (1966) *On the Accuracy of Economic Observations* (Princeton, NJ: Princeton University Press).

Murray, Charles (1988) *In Pursuit: Of Happiness and Good Government* (New York: Simon and Schuster).

Musgrave, Richard and Peggy Musgrave (1989) *Public Finance in Theory and Practice* (5th ed.) (New York: McGraw-Hill).

Nalebuff, Barry (1988) "Puzzles," *Journal of Economic Perspectives*, vol. 2, no. 4 (Fall, 1988), pp. 183-4.

Nathan, Robert (1987) *Social Science: Uses and Misuses* (New York: Basic Books).

Neumann, John von and Oscar Morgenstern (1947) *Theory of Games and Economic Behavior* (Princeton, NJ: Princeton University Press).

Neustadt, Richard and Fineberg, Harvey (1978) *The Swine Flu Affair: Decision-making on a Slippery Disease* (Washington, D.C.: HEW).

Nuclear Regulatory Commission (1975) *Reactor Safety Study: An Assessment of Accident Risks in U.S. Commercial Nuclear Power Plants* (Washington, D.C.: NRC-(WASH-1400)).

Nutter, G. Warren (1983) *Political Economy and Freedom* (Indianapolis, IN: Liberty Press).

O'Driscoll, Gerald P. (1977) *Economics As A Coordination Problem* (Kansas City: Sheed, Andrews, and McNeel).

Olsen, Mancur (1965) *The Logic of Collective Action* (Cambridge: Harvard University Press).

---------- (1982) *The Rise and Decline of Nations* (New Haven: Yale University Press).

Otway, H. J. (1971) *Risk Versus Benefit: Solution or Dream?* (Los Alamos, NM: Western Interstate Nuclear Board).

Parry, G. W. and P. W. Winter (1981) "Characterization and Evolution of Uncertainty in Probabilistic Risk Assessment," *Nuclear Safety*, vol. 22, no. 1.

Pasour, E. C. (1988) "Economic Efficiency and Public Policy," in Block and Rockwell (eds.), pp. 110-24.

Perrow, Charles (1984) *Normal Accidents: Living With High Risk Technologies* (New York: Basic Books).

Phelps, Edmund S. (1985) *Political Economy: An Introductory Text* (New York: W. W. Norton and Company).

Philbrook, Clarence (1953) " 'Realism' in Policy Espousal," *American Economic Review*, vol. 43 (December, 1953), pp. 846-59.

Polanyi, Michael (1962) *Personal Knowledge* (Chicago: University of Chicago Press).

Poole, Robert (1982) "Toward Safer Skies," in *Instead of Regulation* (Lexington, MA: Heath and Company-Lexington Books), pp. 207-38.

Popper, Sir Karl (1957) *The Poverty of Historicism* (New York: Harper Torchbooks).

---------- (1965) *Logic of Scientific Discovery* (New York: Harper and Row).

---------- (1972) "Conjectural Knowledge: My Solution to the Problem of Induction," in *Objective Knowledge: An Evolutionary Approach* (Oxford: Oxford Press), pp. 1-32.

Posner, Richard (1972) *The Economic Analysis of Law* (Boston: Little, Brown and Company).

Radnitzky, Gerard (1987) "Cost-Benefit Thinking in the Methodology of Research: The 'Economic Approach' Applied to Key Problems of the Philosophy of Science," in *Economic Imperialism,* Gerard Radnitzky and Peter Bernholf (eds.) (New York: Paragon House).

Ramsey, Frank (1931) *The Foundations of Mathematics and Other Essays* (New York: Harcourt Brace).

Rave, William (1977) *An Anatomy of Risk* (New York: John Wiley).

Rifkin, Jeremy (1983) *Algeny* (New York: Viking Press).

Rizzo, Mario (1978) "Praxeology and Econometrics: A Critique of Positivist Economics," in Spadaro (ed.), pp. 40-56.

---------- (ed.) (1979) *Time, Uncertainty, and Disequilibrium* (New York: Lexington Books/D.C. Heath).

---------- (1982) "Mises and Lakatos: A Reformulation of Austrian Methodology," in Kirzner (ed.), pp. 53-74.

Robbins, Lionel (1984) *An Essay on the Nature and Significance of Economic Science* (New York: New York University Press).

Roberts, Leslie (1989) "Alar: The Numbers Game," *Science,* vol. 243 (March 17, 1989), pp. 1430.

Rosencranz, Roger D. (1977) *Inferential Method and Decision: Towards A Bayesian Philosophy of Science* (Boston: D. Reidel Publishers).

Rothbard, Murray (1956) "Towards A Reconstruction of Utility and Welfare Economics," in *On Freedom and Free Enterprise: Essays in Honor of Ludwig von Mises,* Hans Sennholz (ed.) (Princeton: D. Van Nostrand). (Reprinted 1977 by the Center for Libertarian Studies).

---------- (1970) *Man, Economy, and State* (Los Angeles: Nash Publishing Company).

---------- (1979) *Individualism and the Philosophy of the Social Sciences* (San Francisco: Cato Institute).

---------- (1982) "Law, Property Rights, and Air Pollution," *Cato Journal,* vol. 2, no. 1, pp. 55-100.

Rothwell, J. Dan (1982) *Telling It Like It Isn't* (Englewood Cliffs, NJ: Prentice Hall).

Rowley, Charles, Robert Tollison, and Gordon Tullock (1988) *The Political Economy of Rent Seeking* (Norwell, MA: Kluwer Academic Publishers).

Russell, Milton and Michael Gruber (1987). "Risk Assessment in Environmental Policy-Making," *Science,* vol. 236 (April 17, 1987), pp. 286-90.

Samuelson, Paul A. (1955) "Diagrammatic Exposition of the Pure Theory of Public Expenditure," in Breit and Hochman (1971), pp. 538-46.

---------- (1974) *The Foundations of Economic Analysis* (New York: Atheneum).

---------- (1982) *The Collected Scientific Papers of Paul A. Samuelson,* J. E. Stiglitz (ed.) (Cambridge, MA: MIT Press).

Sauer, Gerald (1982) "Imposed Risk Controversies: A Critical Analysis," *Cato Journal,* vol. 2, no. 1, pp. 231-50.

Savage, L. J. (1965) *The Foundations of Statistics* (New York: John Wiley).

Schoenbaum, Stephen, Barbara McNeil, and Joel Kavet (1976) "The Swine-Influenza Decision," *New England Journal of Medicine*, vol. 295, no. 14 (September 30, 1976), pp. 759-64.

Schulman, J. L. (1975) "Immunology and Influenza," in Edwin Kilbourne (ed.), pp. 373-93.

Schumpeter, Joseph (1954) *History of Economic Analysis* (Oxford: Oxford University Press).

---------- (1976) *Capitalism, Socialism and Democracy* (New York, NY: Harper Torchbooks).

Schwartz, Barry (1986) *The Battle for Human Nature: Science, Morality, and Human Life* (New York: W. W. Norton).

Schwing, C. and W. Albers (1980) *Societal Risk Assessment: How Safe Is Safe Enough?* (New York: Plenum Press).

Semmons, John and Dianne Kresich (1989) "What If Everything We Know About Safety Is Wrong?" *Liberty*, vol. 2, no. 4 (March, 1989), pp. 37-43.

Shackle, G.L.S. (1969) *Decision, Order, and Time* (Cambridge: Cambridge University Press).

---------- (1972) *Epistemics and Economics* (Cambridge: Cambridge University Press).

Siffin, William B. (1981) "Bureaucracy, Entrepreneurship, and the Barrier Islands," *Cato Journal*, vol. 1, no. 1 (Spring, 1981), pp. 293-311.

Silverstein, Arthur (1981) *Pure Politics and Impure Science: The Swine Flu Affair* (Baltimore: Johns Hopkins University Press).

Simon, Herbert (1976) *Administrative Behavior* (New York: Free Press).

Simpson, David (1983) "Joseph Schumpeter and the Austrian School of Economics," *Journal of Economic Studies*, vol. 4, no. 4, pp. 15-27.

Slovic, Paul (1987) "Perception of Risk," *Science*, vol. 236, April 17, pp. 280-85.

Smith, Merrit Roe (1986) "Technology, Industrialization, and the Idea of Progress in America," in *Responsible Science: The Impact of Technology on Society* (Nobel Conference XXI). (San Francisco: Harper and Row).

Sowell, Thomas (1980) *Knowledge and Decisions* (New York: Basic Books).

---------- (1987) *A Conflict of Visions* (New York: William Morrow).

Spadaro, Louis (ed.) (1979) *New Directions in Austrian Economics* (Kansas City: Sheed, Andrews, and McNeel).

Spengler, Joseph and William Allen (ed.) (1962) *Essays in Economic Thought: Aristotle to Marshall* (Chicago: Rand McNally and Company).

Spiegel, Henry W. and Warren J. Samuels (eds.) (1984) *Political Economy and Public Policy*, vol. 1 (Greenwich, CT: Jai Press).

Stigler, George (1962). "The Economics of Carl Menger," in Spengler and Allen (eds.), pp. 656-71.

---------- (1968) *History of Production and Distribution Theories* (New York: Agathon Press).

Szasz, Thomas (1984) *The Therapeutic State* (New York: Prometheus Books).

Tracey, Michael (1978) "Human Nutrition," in *The Encyclopedia of Ignorance* (New York: Pocket Books).

Tucker, William (1982) *Progress and Privilege: America in the Age of Environmentalism* (Garden City, NY: Anchor Books).

Viscusi, W. and Wesley Magat (1987) *Learning About Risk* (Cambridge, MA: Harvard University Press).

Ward, Benjamin (1979). *The Ideal Worlds of Economics* (New York: Basic Books).

Weaver, Richard M. (1948) *Ideas Have Consequences* (Chicago: University of Chicago Press).

Weber, Max (1978) *Economy and Society,* Guenther Roth and Claus Wittich (eds.). (Berkeley: University of California Press).

Webster, Robert, and Graeme W. Laver (1975) "Antigenic Variation of Influenza Viruses," in Edwin Kilbourne (ed.), pp. 269-314.

Weiser, Friedrich von (1971) *Social Value* (New York: Augustus Kelley).

Whelan, Elizabeth (1985) *Toxic Terror* (Ottowa, IL: Jameson Books).

White, Lawrence (1977) *The Methodology of the Austrian School* (New York: Center for Libertarian Studies).

Whittington, Dale and Duncan MacRae (1986) "The Issue of Standing in Cost Benefit Analysis," *Journal of Policy Analysis and Management,* vol. 5, no. 4.

Wildavsky, Aaron (1987) *Searching for Safety* (New Brunswick, NJ: Transaction Books/Rutgers University).

Wildavsky, Aaron and Mary Douglas (1981) *Risk and Culture* (Berkeley: University of California Press).

Williams, Alan (1977) "Cost-Benefit Analysis: Bastard Science and/or Insidious Poison in the Body Politik?" in Haveman and Margolis (eds.), pp. 519-45.

Wilson, Richard and E. A. C. Crouch (1987) "Risk Assessment and Comparisons: An Introduction," *Science,* vol. 236 (April 17, 1987), pp. 267-70.

Yeager, Leland P. (1980) "Pareto Optimality and Social Decision," *Journal of Libertarian Studies,* vol. 3, no. 4 (Fall, 1980), pp. 3-10.

---------- (1989) "Utility and the Social Welfare Function," in Block and Rockwell (eds.), pp. 175-91.

Zuesse, Eric (1983) "Love Canal: The Truth Seeps Out," in *Reason* (February, 1983), pp. 9-15.

Index